INSTITUTE OF GEOLOGICAL SCIENCES

Natural Environment Research Council

Mineral Assessment

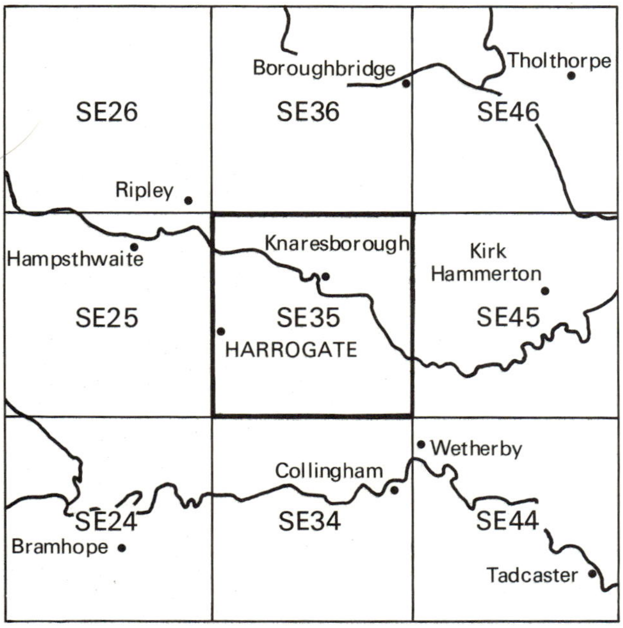

The sand and gravel resources of the country east of Harrogate, North Yorkshire

Description of 1 : 25 000 resource sheet SE 35

D. L. Dundas

Contributor
A. H. Cooper

ISBN 0 11 884170 X*

London Her Majesty's Stationery Office 1981

The first twelve reports on the assessment of British sand and gravel resources appeared in the Report series of the Institute of Geological Sciences as a subseries.

Report 13 and subsequent reports appear as Mineral Assessment Reports of the Institute.

Details of published reports appear at the end of this Report.

Any enquiries concerning this report may be addressed to Head, Industrial Minerals Assessment Unit, Institute of Geological Sciences, Keyworth, Nottingham NG12 5GG.

PREFACE

National resources of many industrial minerals may seem so large that stocktaking appears unnecessary, but the demand for minerals and for land for all purposes is intensifying and it has become increasingly clear in recent years that regional assessments of the resources of these minerals should be undertaken. The publication of information about the quantity and quality of deposits over large areas is intended to provide a comprehensive factual background against which planning decisions can be made.

Sand and gravel, considered together as naturally occurring aggregate, was selected as the bulk mineral demanding the most urgent attention, initially in the south-east of England, where about half the national output is won and very few sources of alternative aggregates are available. Following a short feasibility project, initiated in 1966 by the Ministry of Land and Natural Resources, the Industrial Minerals Assessment Unit (formerly the Mineral Assessment Unit) began systematic surveys in 1968. The work is now being financed by the Department of the Environment and is being undertaken with the cooperation of the Sand and Gravel Association of Great Britain.

This Report describes the resources of sand and gravel of 100 km^2 of country east of Harrogate, North Yorkshire, shown on the accompanying 1 : 25 000 resource map. The survey was conducted in 1978–79 by Mr. D. A. Abraham, Mr D. L. Dundas, Mr J. R. A. Giles and Mr D. Thomas, under the supervision of Mr D. Price. The work is based on six-inch and 1 : 10 000-scale geological surveys carried out in 1976–78 by Mr I. C. Burgess and by Mr A. H. Cooper (who has contributed the description of the geology of the district), of the Yorkshire and East Midlands Unit, to be published on the 1 : 50 000 New-Series Sheet 62 (Harrogate).

Mr W. N. Pierce (Land Agent) negotiated access to land for drilling. The ready cooperation of landowners, tenants, and members of the extractive industry, particularly Tilling Construction Services Ltd, is gratefully acknowledged.

G. M. Brown
Director

Institute of Geological Sciences
Exhibition Road
London SW7 2DE

1 December 1980

CONTENTS

FIGURES

MAP

TABLES

The sand and gravel resources of the country east of Harrogate, North Yorkshire

Description of 1 : 25 000 resource sheet SE 35

D. L. DUNDAS

SUMMARY

The assessment of the sand and gravel resources of the district east of Harrogate is based on geological maps of the Institute of Geological Sciences, pre-existing borehole information, records made available by the sand and gravel industry, and on 38 boreholes drilled for the Industrial Minerals Assessment Unit.

All deposits in the area which might be potentially workable for sand and gravel have been investigated and a simple statistical method has been used to estimate their volume. When feasible, the volume estimates are given at the symmetrical 95 per cent probability level.

The assessed area on the 1 : 25 000 map is divided into three resource blocks, containing between 9.5 and 20.3 km^2 of sand and gravel. For each block the geology of the deposits is described and the mineral-bearing area, the mean thicknesses of overburden and mineral and the mean gradings are stated. Detailed borehole data are also given. The geology, the position of the boreholes and the outlines of the resource blocks are shown on the accompanying map.

Bibliographic reference

DUNDAS, D. L. 1981. The sand and gravel resources of the country east of Harrogate, North Yorkshire. Description of 1 : 25 000 resource sheet SE 35. *Miner. Assess. Rep. Inst. Geol. Sci.*, No. 70.

Author and contributor

D. L. Dundas, MA
Institute of Geological Sciences
Keyworth, Nottingham NG12 5GG

A. H. Cooper, BSc
Institute of Geological Sciences
Ring Road Halton, Leeds LS15 8TQ

INTRODUCTION

The survey is concerned with the estimation of resources, which include deposits that are not currently exploitable but have a foreseeable use, rather than reserves, which can be assessed only in the light of current, locally prevailing, economic considerations. Clearly, both the economic and the social factors used to decide whether a deposit may be workable in the future cannot be predicted; they are likely to change with time. Deposits not currently economically workable may be exploited as demand increases, as higher grade or alternative materials become scarce, or as improved processing techniques are applied to them. The improved knowledge of the main physical properties of the resource and their variability, which this survey seeks to provide, will add significantly to the factual background against which planning policies can be decided (Archer, 1969; Thurrell, 1971; Harris and others, 1974).

The survey provides information at the 'indicated' level "for which tonnage and grade are computed partly from specific measurements, samples or production data and partly from projection for a reasonable distance on geologic evidence. The sites available for inspection, measurement, and sampling are too widely or otherwise inappropriately spaced to permit the mineral bodies to be outlined completely or the grade established throughout" (Bureau of Mines and Geological Survey, 1948, p. 15).

It follows that the whereabouts of reserves must still be established and their size and quality proved by the customary detailed exploration and evaluation undertaken by the industry. However, the information provided by this survey should assist in the selection of the best targets for such further work. The following arbitrary physical criteria have been adopted:

a The deposit should average at least 1 m in thickness.
b The ratio of overburden to sand and gravel should be not more than 3 : 1.
c The proportion of fines (particles passing the No. 240 mesh B.S. sieve, about $\frac{1}{16}$ mm) should not exceed 40 per cent.
d The deposit must lie within 25 m of the surface, this being taken as the likely maximum working depth under most circumstances. It follows from the second criterion that boreholes are drilled no deeper than 18 m if no sand and gravel has been proved.

A deposit of sand and gravel that broadly meets these criteria is regarded as 'potentially workable' and is described and assessed as 'mineral' in this report. As the assessment is at the indicated level, parts of such a deposit may not satisfy all the criteria.

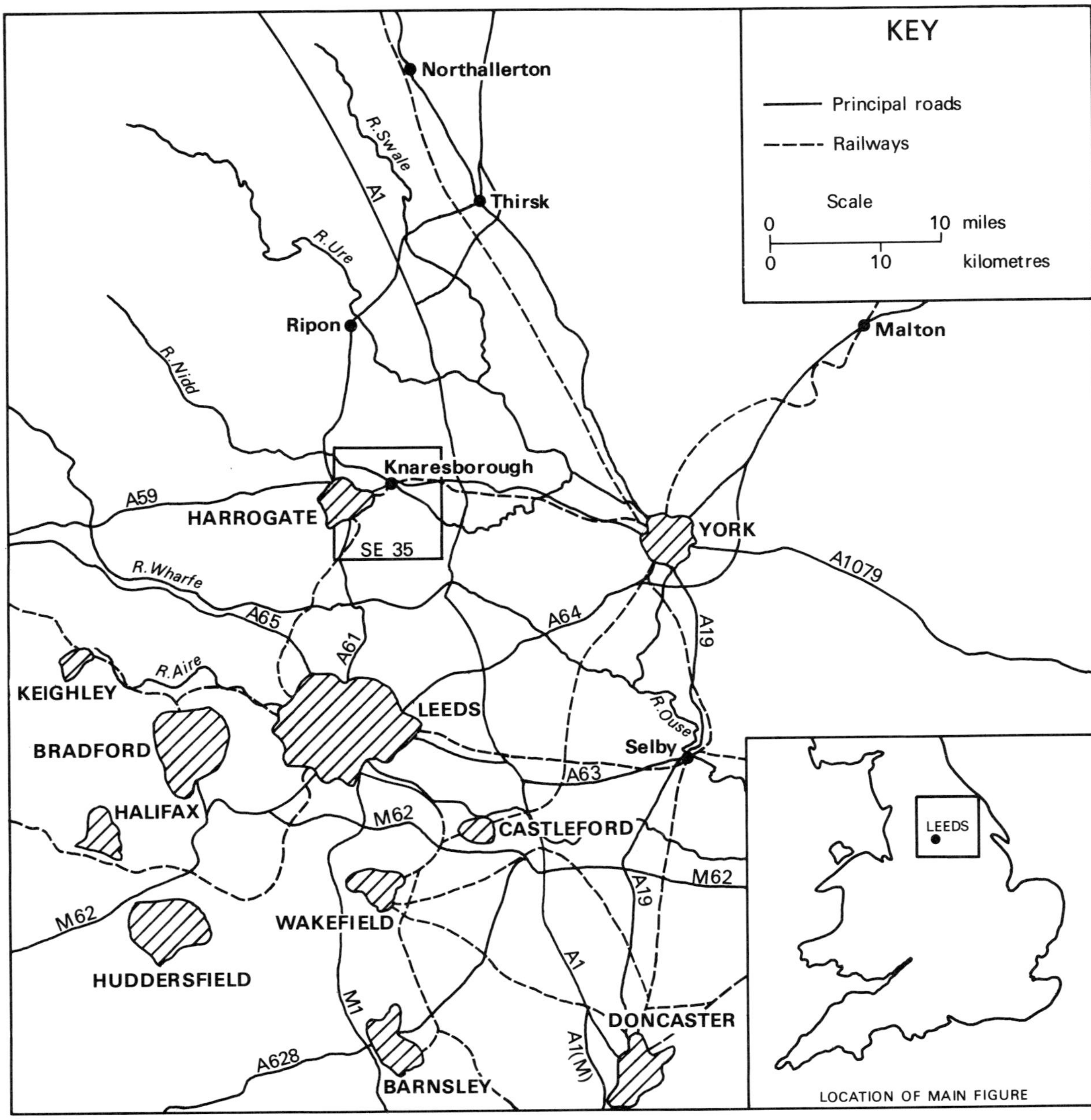

Figure 1 Map showing the location of sheet SE 35.

For the particular needs of assessing sand and gravel resources, a grain-size classification based on the geometric scale $\frac{1}{16}$ mm, $\frac{1}{4}$ mm, 1 mm, 4 mm, 16 mm, 64 mm has been adopted. The boundaries between fines (that is, the clay and silt fractions) and sand, and between sand and gravel material, are placed at $\frac{1}{16}$ mm and 4 mm respectively (see Appendix C).

The volume and other characteristics are assessed within resource blocks, each of which, ideally, contains approximately 10 km^2 of sand and gravel. No account is taken of any features, such as roads, railways, villages or land of high agricultural or landscape value, that might preclude the exploitation of sand and gravel, although urban areas are not assessed. The estimated total volume therefore bears no simple relationship to the amount that could be extracted in practice.

It must be emphasised that the assessment applies to the resource block as a whole. Valid conclusions cannot be drawn about the mineral in parts of a block, except in the immediate vicinity of the actual sample points.

DESCRIPTION OF THE DISTRICT

GENERAL

The area assessed on this resource sheet (for location, see Figure 1) amounts to 48.7 km^2, of which 16.4 km^2 is mineral-bearing. The urban areas of Harrogate (12.2 km^2) and Knaresborough (3.4 km^2), and the working and worked-out sand and gravel areas (1.65 km^2) north, north-east and south-east of Knaresborough have not been assessed. To date (1979) an estimated volume of about 8 million m^3 of sand and gravel has been extracted from the district.

Harrogate lies astride the western margin of the resource sheet, whilst the market town of Knaresborough is just north of its centre on the left (north-east) bank of the River Nidd. The York–Harrogate road (A59) and railway run in an east–west direction across the northern half of the area.

TOPOGRAPHY

The district bestrides the western margin of the Vale of York. The higher ground in its western half, which rises,

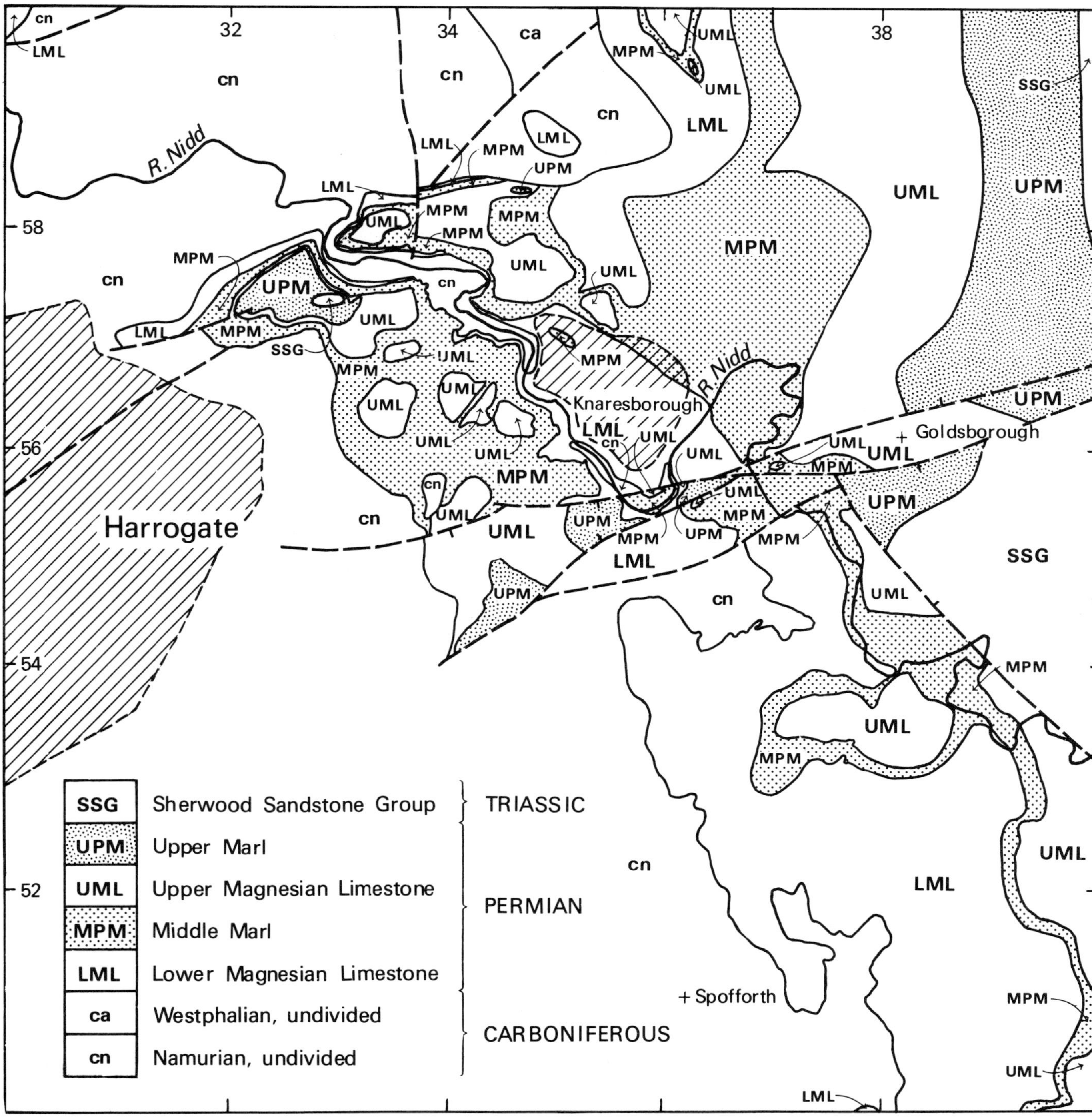

Figure 2 Solid geology.

south of Pannal, to a maximum elevation of 160 m above OD, is part of the eastern border of the Pennines. Harrogate is built on an upland area with moderate relief that varies in elevation between 75 and 140 m above OD; this upland, over which the superficial deposits are generally thin (usually less than 10 m thick), is transected by the River Nidd, which has cut a gorge more than 45 m deep just west of Knaresborough. The Nidd emerges from its gorge in the south-eastern outskirts of Knaresborough and then flows south-eastwards in a sinuous channel which is generally flanked by high banks of alluvium and terrace deposits. The southern half of the district is drained by Crimple Beck, which flows south-eastwards in a relatively mature valley before swinging northwards east of Spofforth to reach its confluence with the Nidd east of Little Ribston.

North-east of a line through Scotton, Scriven and Little Ribston the relief is relatively subdued and the bedrock is for the most part concealed by superficial deposits that are usually between 10 and 25 m thick. Between Scriven and Coneythorpe, Till, with intercalated lenticular bodies of Glacial Sand and Gravel, forms undulating ground between 40 and 60 m above OD. Further south, between Goldsborough [380 560] and Ribston Park [394 543], is an area of unusually low relief underlain by lacustrine clay. Brown Moor [385 585] is a similarly low-lying area, likewise underlain by lacustrine deposits.

GEOLOGY

The Solid (Figure 2) and Drift sequences found at and near the surface in the district are listed in Table 1, and the relationships between them are illustrated in Figure 3.

SOLID

Namurian and Westphalian The Carboniferous rocks comprise a sequence of siltstones and mudstones with

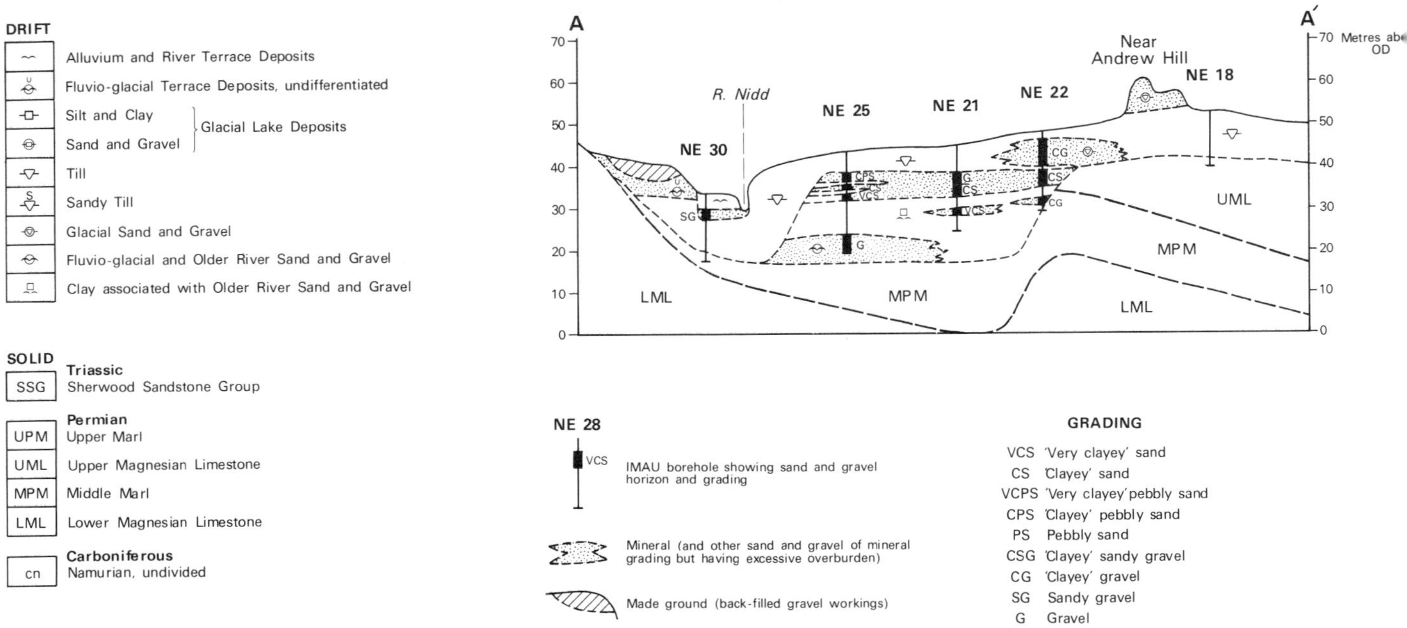

Figure 3 Cross-sections showing the relationships of the Drift deposits. The lines of section are indicated on the resource map.

Table 1 Geological succession

DRIFT	
Quaternary	Peat
	Blown Sand
	Alluvium
	River Terrace Deposits, undifferentiated
	Fluvio-glacial Terrace Deposits, undifferentiated
	Glacial Lake Deposits
	Laminated Clay closely associated with Glacial Deposits
	Glacial Sand and Gravel
	Till and Sandy Till
	Clay associated with Fluvio-glacial and Older River Sand and Gravel
	Fluvio-glacial and Older River Sand and Gravel
SOLID	
Triassic	Sherwood Sandstone Group
Permian	Upper Marl
	Upper Magnesian Limestone
	Middle Marl
	Lower Magnesian Limestone
Carboniferous	Westphalian, undivided
	Namurian, undivided

interbedded sandstones and gritstones, in total more than 800 m thick. At several horizons the mudstones yield goniatites and other marine fossils which show that the strata present range in age from early Namurian (Millstone Grit) to early Westphalian (Coal Measures). The oldest rocks underlie the town of Harrogate, where they form the core of the Harrogate Anticline. The axis of this fold plunges north-eastwards, so that successively younger strata crop out to the north and to the south-east of the town. The youngest (Westphalian) strata are preserved only in the Farnham area, where they partly floor the valley occupied by the Farnham Gravel Quarry [350 600] (Burgess and Cooper, 1981).

Lower Magnesian Limestone This formation dips to the east and forms a faulted escarpment that runs through Rabbit Hill [356 597], the southern part of Knaresborough [353 565], and North Deighton [390 515]. The Lower Magnesian Limestone rests unconformably on the Carboniferous rocks, and variations in its thickness (0–45 m) are largely due to irregularities of the surface of the latter; however, west of Knaresborough, depositional thinning of the Limestone is evident. The Lower Magnesian Limestone consists mainly of thinly-bedded to massive, dolomite and dolomitic limestone; many beds are oolitic and vughs are common; large-scale cross-bedding abounds in the upper part of the formation.

Middle Marl Reddish brown, thinly-bedded calcareous mudstones comprise most of the Middle Marl, but it also contains interbedded gypsum and anhydrite, mainly at depth. There are few exposures of the Marl, which generally forms a low-lying Drift-covered tract parallel to the outcrop of the Lower Magnesian Limestone; west of Knaresborough it forms a large outlier. The thickness of the Marl is variable (0–25 m); the variation is due in part to differential solution of the interbedded evaporites. West and south of Knaresborough, the Middle Marl overlaps the Lower Magnesian Limestone onto the Carboniferous rocks and wedges out to the south-west.

Upper Magnesian Limestone Around Hopewell House [373 587] the Upper Magnesian Limestone is concealed by Drift, but near Goldsborough [378 555], Little Ribston [382 535] and Kirk Deighton [398 505] it forms a prominent scarp. It also forms small outliers at Gibbet Hill [358 599] and west of Knaresborough. It is between 7 and 22 m thick, and composed largely of thinly-bedded, dolomitic limestone with vugs. South of Knaresborough the Limestone overlaps the Middle Marl onto the Carboniferous rocks.

Upper Marl There are no exposures of the Upper Marl in the east of the district, where virtually all of the outcrop is obscured by Drift. However, to the west and south of

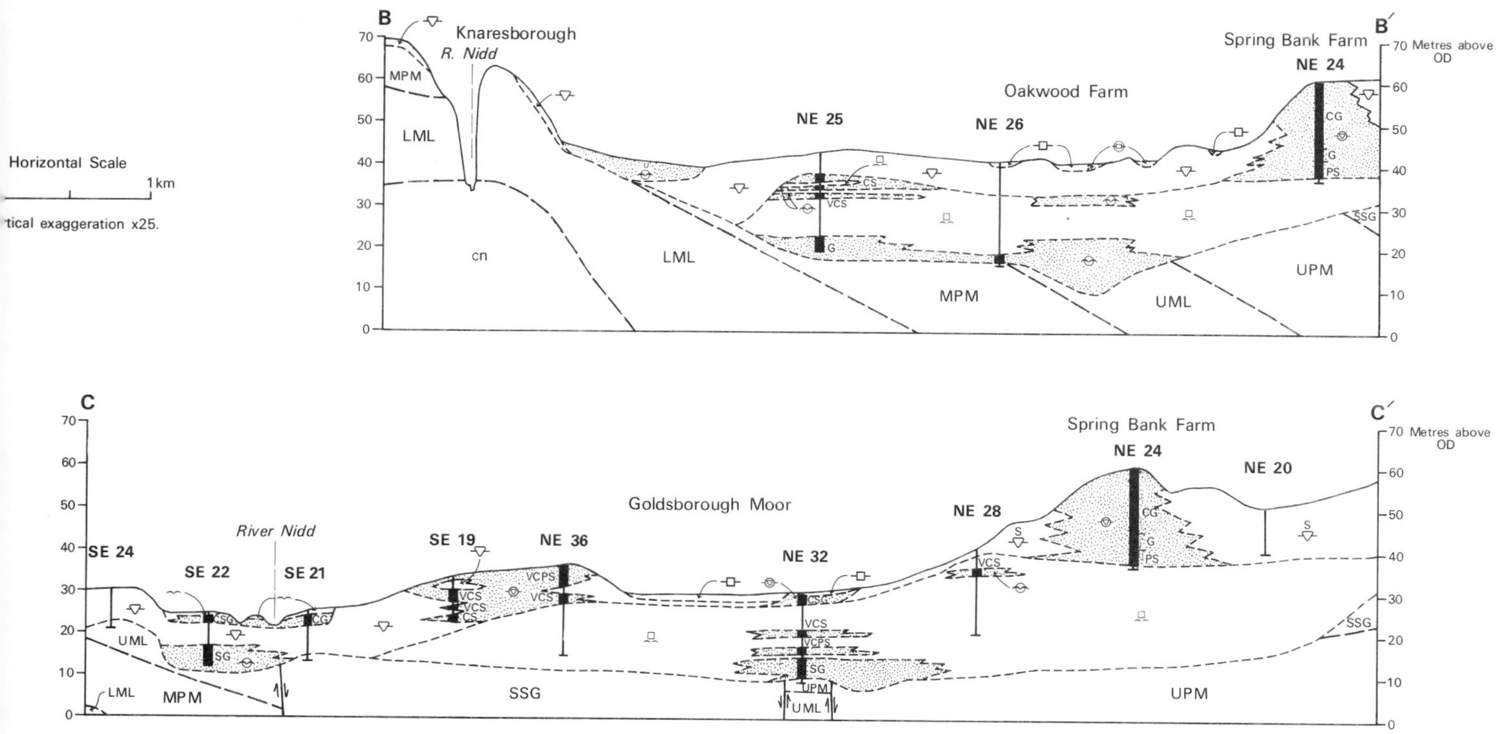

Knaresborough the Upper Marl is exposed in outliers. Boreholes east of the sheet area prove 21 to 23 m of reddish brown silty mudstone with gypsum and anhydrite. A few thin sandstones are present, mainly near the top of the formation, where there is a passage by alternation into the overlying Sherwood Sandstone.

Sherwood Sandstone Group This Group consists of reddish brown, fine- and medium-grained sandstones, the lowermost 50 metres of which subcrop beneath Drift along the eastern margin of the district south-east of Goldsborough. A small, poorly exposed outlier is present in the west near Bilton Hall [330 575].

DRIFT

Fluvio-glacial and Older River Sand and Gravel This deposit predates the Till and overlies the bedrock in the buried valley that runs south-eastwards from Farnham [350 600] to east of Goldsborough (Figure 4). Its deposits are considered to be fluvioglacial outwash gravels that partially fill a former river channel (Johnson, 1974).

Sand and gravel, up to 22.5 m thick in the area of Knaresborough Golf Course [356 590] (boreholes NE 5 and NE 6), occupies the buried valley and is now concealed by a cover of Till. North of Knaresborough, the flanks of the valley are fairly steep and well defined, but farther east, near Goldsborough, the valley broadens and the positions of the northern and southern margins of the concealed sand and gravel deposits it contains are uncertain. Where the valley widens, east of Hall Farm [372 579], the gravel thins and becomes interdigitated with sand and laminated clay; the proportion of clay increases towards the east.

Clay associated with Fluvio-glacial and Older River Sand and Gravel Clay, usually laminated and up to 14 m thick, is associated with the Fluvio-glacial and Older River Sand and Gravel and fills the eastern part of the buried valley mentioned above. This clay is generally overlain by Till, but at Goldsborough Park [386 558] it is concealed beneath lithologically similar Glacial Lake Deposits; north of Little Ribston [385 545] River Terrace Deposits overlie the clay; farther south [385 522] it is overlain by Fluvio-glacial Terrace Deposits.

Till and Sandy Till Deposits of Till and Sandy Till, up to 20 m thick, are widespread. The matrix of the tills varies from clay to sandy clay, and is commonly silty. Pebbles, cobbles and boulders in the tills consist predominantly of Carboniferous sandstone, although Carboniferous limestone is locally abundant. Fragments of Permian dolomitic limestone are scarce except near the outcrops of the Lower and Upper Magnesian Limestones. Lenses and discontinuous beds of laminated clay, sand and gravel are present within the tills.

Laminated Clay closely associated with Glacial Deposits Small patches of laminated clay with silt and sand beds occur within the till sequence. A borehole (NW 23) north of Scotton proved 7.8 m of these deposits. The clays are associated with Till, Sandy Till and Glacial Sand and Gravel; they postdate the Clay associated with Fluvio-glacial and Older River Sand and Gravel but predate the Silt and Clay, Glacial Lake Deposits.

Glacial Sand and Gravel The deposits of Glacial Sand and Gravel vary considerably in their composition, thickness, topographic expression and their relationships to the till. They vary from fine- to medium-grained clayey sand with clay laminae to coarse gravel with boulders in a clayey sand matrix. The gravel is composed largely of pebbles of Carboniferous sandstone and limestone. The deposits occur above, within and below the till, and were probably laid down by water on top of, adjacent to, or beneath a stagnating ice-sheet. Some of the clayey gravels, however, are probably of morainic origin.

Fluvio-glacial Terrace Deposits Adjacent to the River Nidd small patches of high terrace deposits, mainly consisting of clayey gravel, extend across the district from the north-west to the south-east corner. These terraces lie about 15 m above the present level of the River Nidd and descend in elevation from about +65 m above OD in the north-west to about +45 m south-east of

Knaresborough; here they have been worked in the past on both sides of the river near Grimbald Bridge [362 562]. From Knaresborough the terraces grade down south-eastwards to the Kirk Deighton Channel at about + 30 m (Edwards and others, 1950), west of Kirk Deighton [395 403].

Glacial Lake Deposits Clays, silts, sands and gravels of lacustrine origin form extensive flat expanses occupying the low ground in the north-east of the district. Stiff, grey to brown, stoneless clay with laminae of silt and fine sand forms the bulk of the lacustrine deposits. At Brown Moor [384 586] the deposit is thin and overlies till, as it does at Goldsborough Fields [392 565], but at Goldsborough Park, where till is locally absent, it rests on Clay associated with Fluvio-glacial and Older River Sand and Gravel. Thin sands are interbedded with the lacustrine clays and are common on Brown Moor where they form a marginal facies to the lacustrine clays. The edges of these marginal sands are commonly gravelly and represent beach deposits.

River Terrace Deposits Several terraces flank the River Nidd and Crimple Beck at elevations below that of the Fluvio-glacial Terrace Deposits. The higher of these terraces are generally composed of sand and gravel and the lower ones of silt and clay.

Alluvium Clay and silt with local patches of sand form the river alluvium, which is generally less than 4 m thick. Sand and gravel are commonly present at the base, and were probably laid down as channel-lag deposits.

Numerous ill-drained hollows contain alluvial clay and silt, commonly associated with peat. The majority are situated in areas mantled by till, and some are typical kettle holes. Other hollows, largely confined to areas underlain by Middle Marl, Upper Magnesian Limestone or Upper Marl, are subsidence features resulting from the removal by solution of evaporites from the marl formations.

Peat Peat, generally less than 2 m thick, occupies many areas of low-lying, ill-drained land. On Brown Moor a large peat-filled depression [386 588] on the lacustrine flat was probably produced by subsidence following the solution of underlying evaporites.

Blown Sand Fine sand up to 1 m thick mantles several west-facing slopes around Brown Moor. This sand has been blown eastwards from the lacustrine deposits and is still actively migrating.

GRADING AND COMPOSITION OF THE SAND AND GRAVEL DEPOSITS

The mean grading of the mineral proved by IMAU boreholes in each deposit in the district is given in Table 2. However, individual mineral units often exhibit gradings that differ widely from the mean grading of the deposit to which they belong. Thus, although the overall mean grading of the *Fluvio-glacial and Older River Sand and Gravel* is fines 11 per cent, sand 59 per cent and gravel 30 per cent, resulting in a classification (Figure 11) as 'clayey' sandy gravel, some mineral units assigned to this deposit grade as 'clayey' or 'very clayey' sand (for example, in boreholes NE 21, 22, 25, 28 and 31), while others – often in the same borehole – grade as gravel or sandy gravel. The wide variation from the mean grading proved for mineral units belonging to this deposit on Block A is illustrated in a mean particle-size distribution diagram (Figure 6).

The *Glacial Sand and Gravel* likewise includes deposits that diverge strongly in their gradings from the overall mean. However, the variation in grading found at any one locality is less than it often is in the Fluvio-glacial and Older River Sand and Gravel, in which waste partings are quite common and in which the mineral units separated by these waste partings may show strongly contrasting mean gradings (as in boreholes NE 22, 25 and 32).

The overall mean grading proved in IMAU boreholes for the Glacial Sand and Gravel is fines 19 per cent, sand 46 per cent and gravel 35 per cent; it is thus classified overall as 'clayey' sandy gravel. However, it is not certain that this mean grading, calculated on the basis of data from 9 boreholes, represents a close approximation to the actual overall mean grading, since there are numerous separate small bodies of Glacial Sand and Gravel that have not been sampled. There is some indication in the way of field mapping evidence that some of the smaller bodies are clayey sand. The available data indicate that the mineral in an individual body of Glacial Sand and Gravel is likely to be fairly homogeneous in its grading characteristics and to be composed either of 'clayey' gravel (boreholes NW 23 and 25, NE 22 and 24), occasionally 'clayey' sandy gravel (borehole NE 32), or else of 'very clayey' sand (boreholes NE 36, SE 19) or 'very clayey' pebbly sand (NE 36, SE 16 and 26). It is possible that this apparent twofold division of the mineral bodies into those that consist of 'clayey' gravel, with a mean grading of fines 15 per cent, sand 38 per cent and gravel 47 per cent, and those that consist of 'very clayey' sand, with a mean grading of fines 28 per cent, sand 69 per cent and gravel 3 per cent, may merely reflect the sparsity of the available data, but it is considered more likely to be real and to reflect distinct modes of deposition of the mineral bodies.

Mineral assigned to the *Glacial Lake Deposits* was found in only one IMAU borehole (NE 23), which proved 4.0 m of 'clayey' sand; it is considered unlikely that the limited volume of mineral in this deposit will be commercially attractive in the foreseeable future.

Only three patches of *Fluvio-glacial Terrace Deposits* had IMAU boreholes sited upon them (i.e. boreholes NE 29 and 34, and SE 23). The patch near Grimbald Bridge [362 562] has mostly been worked out and the remaining mineral ('very clayey' pebbly sand) is covered by thick overburden (10.7 m in borehole NE 29). The mineral in the patch south-west of Goldsborough is 'clayey' gravel with a mean grading (in borehole NE 34) of fines 13 per cent, sand 28 per cent and gravel 59 per cent. However, its thinness (1.2 m at this locality) renders it of marginal commercial potential and the underlying mineral in the Older River Sand and Gravel has even less potential, since it consists of 'clayey' and 'very clayey' sand with clay waste partings. The patch lying north-west of North Deighton was found in borehole SE 23 to contain mineral 2.9 m thick with a mean grading of fines 25 per cent, sand 45 per cent and gravel 30 per cent; it is thus classified as a 'very clayey' sandy gravel. This deposit is considered to be the most prospective outcrop of the Fluvio-glacial Terrace Deposits in the district although the extent of the mineral-bearing area in the stratigraphically equivalent patch on the west bank of Crimple Beck, is uncertain.

The mean gradings determined for the mineral of the *Alluvium* and of the *River Terrace Deposits* are similar (Table 2), except that the percentage of fines is somewhat

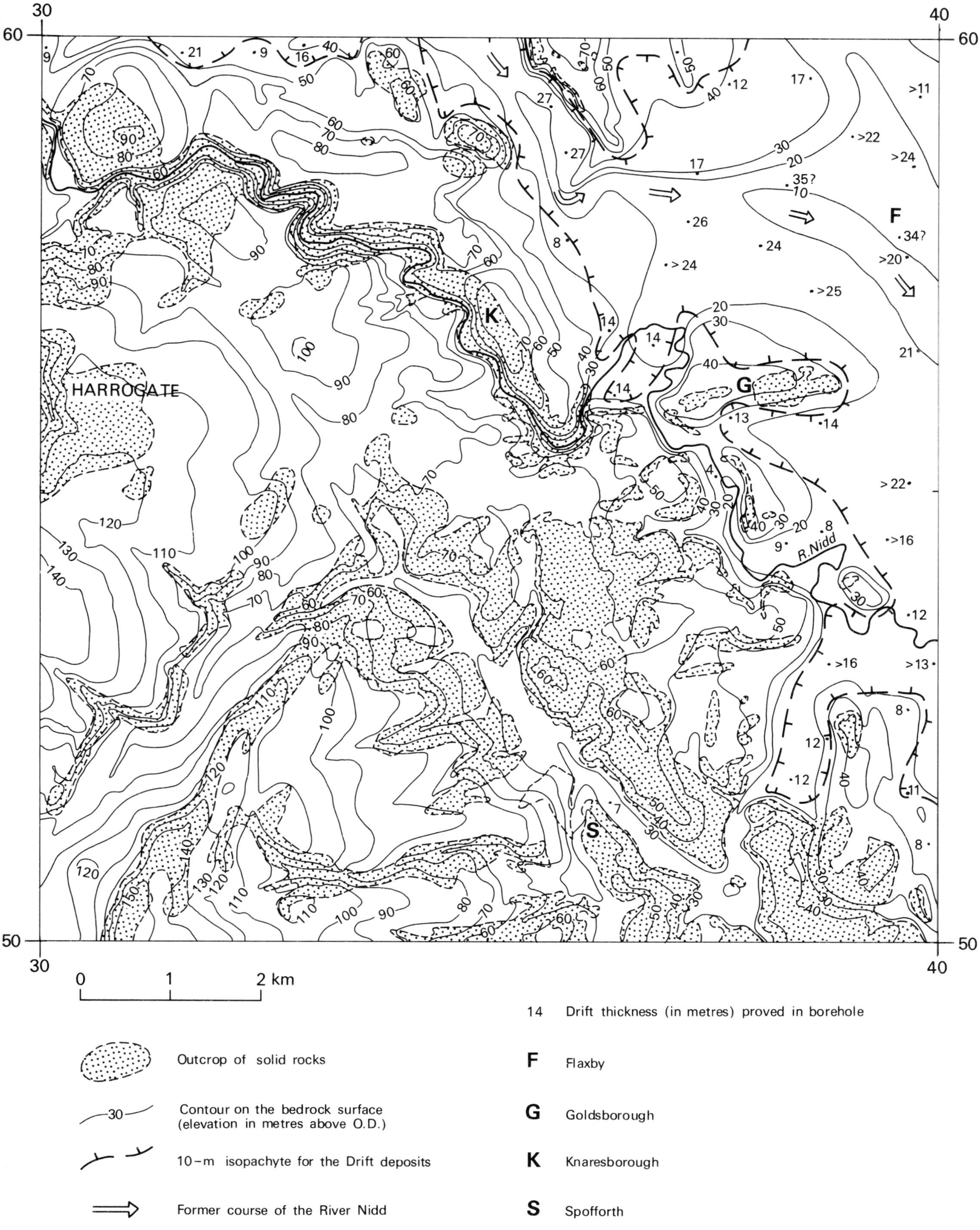

Figure 4 Contour map of the bedrock surface, showing the former course of the River Nidd and selected borehole data on the thickness of the Drift deposits.

Table 2 Mean gradings of the mineral proved in IMAU boreholes, by deposit

Deposit*	Mean grading percentages						
	Fines $-\frac{1}{16}$ mm	Fine sand $+\frac{1}{16}-\frac{1}{4}$ mm	Medium sand $+\frac{1}{4}-1$ mm	Coarse sand $+1-4$ mm	Fine gravel $+4-16$ mm	Coarse gravel $+16-64$ mm	Cobbles $+64$ mm
Fluvio-glacial and Older River Sand and Gravel (9)	11	23	25	11	14	15	1
Glacial Sand and Gravel (9)	19	22	14	10	14	19	2
Fluvio-glacial Terrace Deposits (3)	23	21	25	10	9	11	1
Sand and Gravel, Glacial Lake Deposits (1)	14	77	7	1	1	0	0
River Terrace Deposits (1)	19	13	33	10	8	17	0
Alluvium (3)	10	12	26	9	12	30	0
Mean	16	24	19	10	13	17	1

* The numbers of boreholes upon which the calculations of mean grading are based are given in parentheses.

higher in the latter and that of coarse gravel is distinctly higher in the former. The mineral in both is classified as 'clayey' sandy gravel.

The pebbles found in the Drift deposits of the district are all apparently derived from Upper Palaeozoic rocks and have probably travelled, by a combination of glacial and fluvial transport, fairly short distances, mostly perhaps between 5 and 50 km from the outcrops from which they originate. Nearly all the pebbles belong to one or other of four rock types:

1 Hard, virtually unweathered, grey, brownish grey or black relatively pure limestone, presumably derived from the Viséan Carboniferous Limestone formation that crops out north and west of Skipton (30 km west of Harrogate);

2 Hard, medium- or coarse-grained, beige or light grey sandstone consisting mainly of quartz, but often containing a little feldspar and mica; the pebbles have a low porosity and permeability, and were presumably derived from the Namurian Millstone Grit Series, which crops out in a broad belt of country extending over several hundred square kilometres west and north-west of Knaresborough;

3 Buff, purplish or light greenish-brown siliceous siltstone, likewise presumably derived from the Namurian;

4 Carious, buff-weathered dolomitic limestone, presumably derived from the Permian Magnesian Limestones, which crop out in a belt about five kilometres wide that runs diagonally across the district with a north-north-westerly trend (Figure 2).

Minor elements among the pebbles are vein-quartz, chert and concretionary ironstone (boxstone). Of the major components of the gravels, only the carious dolomitic limestone is likely to lower the quality of the aggregate if present in large amounts. However, high percentages of this component are uncommon, usually occurring only in gravels that directly overlie Magnesian Limestone bedrock. Among the minor components the ferruginous boxstones may be deleterious but their percentage is usually low – between 1 and 3 per cent by weight of the gravel fraction.

Systematic analysis of the lithology of the gravels has not been attempted. The gravels of the Fluvio-glacial and Older River Sand and Gravel are currently being worked north of Knaresborough and the fundamental character of their stone content is unlikely to change markedly within the assessment area. Analysis of selected samples from this deposit (see the records of boreholes NE 25, 26 and 32 for details) shows that pebbles of sandstone and siliceous siltstone, presumably derived from the Millstone Grit Series, together form between 47 and 81 per cent of the gravel fraction; hard non-dolomitic limestone, presumably derived from the Carboniferous Limestone, forms only 3 per cent of the gravel fraction in the sample from borehole NE 32, but 23 per cent in that from borehole NE 25; carious dolomitic Permian limestone forms 11 to 27 per cent whilst vein quartz and chert make up between 4 and 7 per cent and ferruginous boxstone between a trace and 3 per cent of the gravel fraction.

Samples from the Glacial Sand and Gravel deposit at Spring Bank Farm (borehole NE 24) were found to contain the same types of pebbles as were found in the Fluvio-glacial and Older River Sand and Gravel, but the percentage of (Carboniferous) limestone was much higher (56–69 per cent), whilst dolomitic (Permian) limestone was somewhat less common (7–9 per cent). It is suggested that the high proportion of far-travelled but nevertheless mainly subangular Carboniferous Limestone pebbles in this deposit reflects a dominantly englacial mode of transport of the pebbles from their source area some 50 km to the north-west.

The lithology of the fluvial gravels of the Nidd and Crimple Beck valleys has not been analysed quantitatively, but field examination indicates that, as might be expected, the same four main pebble types are present and that Namurian sandstone pebbles predominate.

THE MAP

The sand and gravel resource map is folded into the pocket at the end of this report. The base map is the Ordnance Survey 1 : 25 000 Outline Edition in grey, on which the topography is shown by contours in green, the geological data in black and the mineral resource information in shades of red.

Geological data The geological boundary lines, symbols, etc., shown are taken from the geological map of this area recently surveyed at the scales of 1 : 10 560 and 1 : 10 000. This information was obtained by detailed application of field mapping techniques by the field staff of the Institute's Yorkshire and East Midlands Unit.

Borehole data, which include the stratigraphic relations, thicknesses and mean particle size distribution of the sand and gravel samples collected during the assessment, are also shown. The geological boundaries are regarded as the best interpretation of the information available at the time of the survey.

Mineral resource information The mineral-bearing ground is subdivided into resource blocks (see Appendix A). Within a resource block the mineral is subdivided into areas where it is exposed, that is where overburden averages less than 1 m in thickness, and areas where it is present in continuous or almost continuous spreads beneath overburden.

Areas where bedrock crops out, or where boreholes indicate that sand and gravel is either absent or else present but not potentially workable because of excessive overburden or inadequate thickness (see Introduction), are uncoloured on the map; where appropriate, the relevant criterion is shown. In such cases it has been assumed that mineral is absent except in infrequent and relatively minor patches that can neither be outlined nor assessed quantitatively in the context of this survey. Areas of sand and gravel, or of potentially aggregate-bearing deposits that were not assessed because they occur as small and isolated patches or in urban areas, are indicated by a red stipple.

The extent of the mineral-bearing area is delimited, where feasible, by the mapped geological boundary lines. The whole of this area is taken to be mineral-bearing although it may include some areas in which sand and gravel is absent or not potentially workable. Inferred boundaries have been inserted where sand and gravel beneath cover is interpreted to be not potentially workable or absent. Such boundaries (for which a distinctive symbol is used) are drawn primarily for the purpose of volume estimation. The symbol is intended to convey an approximate location within a likely zone of occurrence rather than to represent the breadth of the zone, its size being limited only by cartographic considerations. For the purpose of measuring areas the centre-line of the symbol is used.

Worked areas and made ground The approximate extent of known sand and gravel workings to Spring 1979 are shown on the map; active and disused workings are indicated, together with area which have been returned to agricultural use and areas partly backfilled with waste from the sand and gravel industry. Workings which have been filled with mixed waste or refuse are shown as 'made ground'.

RESULTS

The statistical results are summarised in Table 6. Fuller grading particulars are shown in Figures 5 to 8, and tabulated in Tables 3 to 5.

Accuracy of results The resources of sand and gravel have been statistically assessed (Table 6) in two resource blocks (A and B), although an inferred assessment was made for part of the mineral in Block A; for the third block, C, an inferred assessment is given. For the two

Table 3 Data from IMAU boreholes: Block A

Borehole	Recorded thickness			Mean grading percentages						
	Mineral m	Over-burden m	Waste partings m	Fines $-\frac{1}{16}$ mm	Fine sand $+\frac{1}{16}-\frac{1}{4}$ mm	Medium sand $+\frac{1}{4}-1$ mm	Coarse sand $+1-4$ mm	Fine gravel $+4-16$ mm	Coarse gravel $+16-64$ mm	Cobbles $+64$ mm
All deposits										
NE 17	absent									
NE 18	absent									
NE 19	absent									
NE 20	absent									
NE 21	7.2	6.5	2.5	13	24	28	8	11	15	1
NE 22	13.2	1.6	2.3	15	24	15	8	13	21	4
NE 23	4.0	0.9	–	14	77	7	1	1	0	0
NE 24	22.4	0.4	–	14	15	13	14	18	24	2
NE 25	8.4+	4.8	10.4	11	30	17	6	12	23	1
NE 26	absent									
NE 27	absent									
NE 28	1.4	4.1	–	23	58	17	2	0	0	0
NE 31	3.7	8.9	–	14	31	48	4	3	0	0
NE 32	5.6	0.5	13.7	17	25	30	7	12	9	0
NE 35	absent									
NE 36	6.4	0.6	2.0	30	46	18	3	2	1	0
SE 19	5.5+	3.0	2.0	23	60	17	0	0	0	0
Mean	7.8	3.1	3.3	16	30	19	8	11	15	1
Fluvio-glacial and Older River Sand and Gravel										
NE 21	7.2	6.5	2.5	13	24	28	8	11	15	1
NE 22	6.9	2.2	1.7	13	36	22	7	8	12	2
NE 25	8.4+	4.8	10.4	11	30	17	6	12	23	1
NE 28	1.4	4.1	–	23	58	17	2	0	0	0
NE 31	3.7	8.9	–	14	31	48	4	3	0	0
NE 32	4.6	13.7	–	6	12	42	10	16	13	1
Mean	5.4	6.7	2.4	12	29	27	7	10	14	1

Table 4 Data from IMAU boreholes: Block B

Borehole	Recorded thickness			Mean grading percentage						
	Mineral	Over-burden	Waste partings	Fines	Fine sand $+\frac{1}{16}-\frac{1}{4}$ mm	Medium sand $+\frac{1}{4}-1$ mm	Coarse sand $+1-4$ mm	Fine gravel $+4-16$ mm	Coarse gravel $+16-64$ mm	Cobbles
	m	m	m	$-\frac{1}{16}$ mm						$+64$ mm
NE 29	3.7	10.7	–	25	16	23	19	10	7	0
NE 30	2.1	4.0	–	5	9	43	9	9	25	0
NE 33	absent									
NE 34	4.2	0.4	3.4	21	32	27	4	5	9	2
SE 16	2.5	4.6	–	27	16	25	17	9	6	0
SE 17	2.2	6.4	–	14	7	15	17	27	19	1
SE 18	absent									
SE 20	absent									
SE 21	2.8	0.7	–	16	18	17	4	11	34	0
SE 22	5.4+	1.0	5.7	7	5	25	28	19	15	1
SE 23	2.9	0.1	–	25	12	25	8	13	17	0
SE 25	4.0	7.5	–	8	1	14	26	32	19	0
SE 28	1.5	3.0	–	6	4	20	20	18	32	0
Mean	3.1	3.8	0.9	16	12	23	16	15	17	1

Table 5 Data from IMAU boreholes: Block C

Borehole	Recorded thickness			Mean grading percentage						
	Mineral	Over-burden	Waste partings	Fines	Fine sand $+\frac{1}{16}-\frac{1}{4}$ mm	Medium sand $+\frac{1}{4}-1$ mm	Coarse sand $+1-4$ mm	Fine gravel $+4-16$ mm	Coarse gravel $+16-64$ mm	Cobbles
	m	m	m	$-\frac{1}{16}$ mm						$+64$ mm
NW 22	absent									
NW 23	3.7	9.7	–	15	8	14	16	20	25	2
NW 24	absent									
NW 25	9.6	2.4	–	17	15	12	10	18	25	3
NW 26	absent									
SE 24	absent									
SE 26	1.0	0.3	–	38	48	12	2	0	0	0
SE 27	absent									
Mean	4.2	4.3	–	19	15	15	12	16	21	2

Table 6 The sand and gravel resources of the country around Knaresborough, North Yorkshire: summary of statistical results

Block	Mineral subdivision†	Area		Mean thickness		Volume of sand and gravel			Mean grading percentage		
		Block	Mineral	Over-burden	Mineral		Limits at the 95% confidence level		Fines	Sand $+\frac{1}{16}$	Gravel
		km²	km²	m	m	$m^3 \times 10^6$	±%	$\pm m^3 \times 10^6$	$-\frac{1}{16}$ mm	-4 mm	$+4$ mm
A	1	20.3	0.7‡	0.9	4.0	>2.0¶	–	–	14	85	1
	2	20.3	2.1‡	–	–	>12.8¶	–	–	–	–	–
	3	20.3	7.8*‡	6.7	5.4	42.1	54	23	12	63	25
B		9.5	6.0§	3.8	3.1	18.6	28	5	16	51	33
C[1]		13.4	0.3¶	–	–	3.0¶	–	–	–	–	–
C[2]		4.7	0.1	–	–	0.2¶	–	–	–	–	–

* Excludes current workings (1979) and worked-out areas totalling about 98 hectares.
† Subdivisions of the mineral in Block A are as follows:
1 Sand and Gravel, Glacial Lake Deposits and Blown Sand.
2 Glacial Sand and Gravel.
3 Fluvio-glacial and Older River Sand and Gravel.
‡ These areas partially overlap.
§ Excludes worked-out area of 27 hectares.
|| Excludes about 47 hectares of deposits that may contain mineral but were not assessed, and about 7 hectares of worked-out ground.
¶ Inferred assessment.

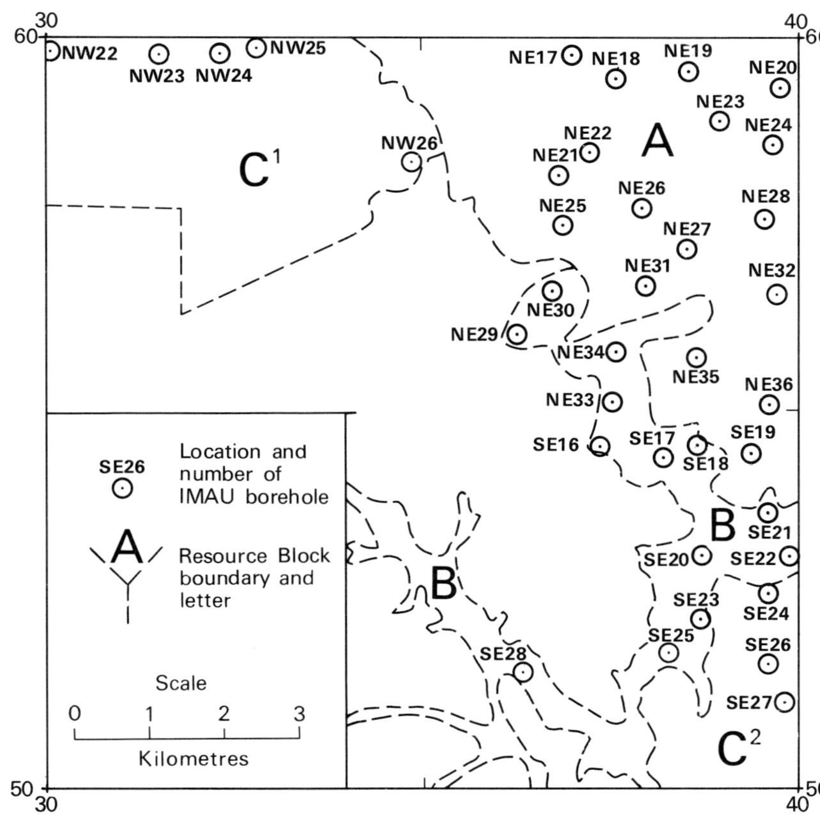

Figure 5 The gradings of individual mineral units and overall mean gradings proved in IMAU boreholes.

For the sake of completeness, sand and gravel units that are 'non-mineral' because of excessive overburden have been included in the columnar sections.

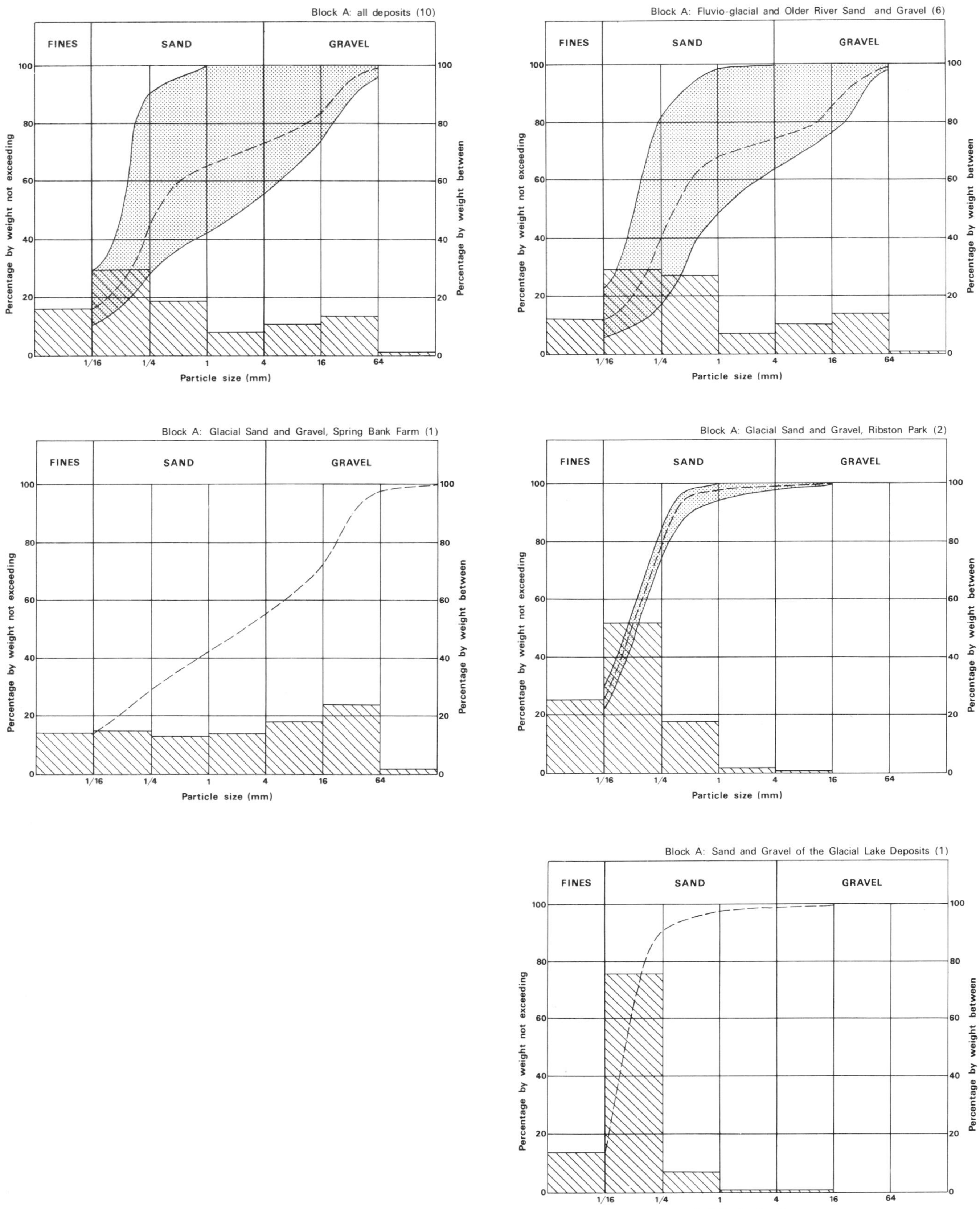

Figure 6 Mean particle-size distribution for the assessed thickness of mineral in Block A and in some of the mineral-bearing deposits within it.

The numbers in parentheses indicate the numbers of IMAU boreholes used to construct the diagrams. The shaded areas indicate the envelope within which the respective deposits fall. The stippled columns indicate the percentages that fall within the respective particle-size ranges.

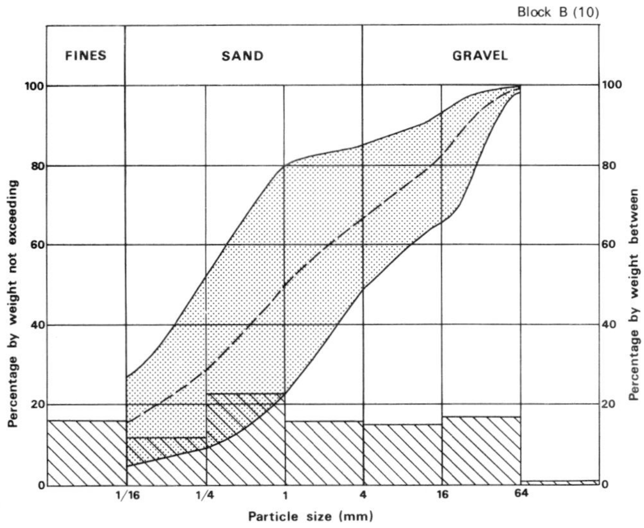

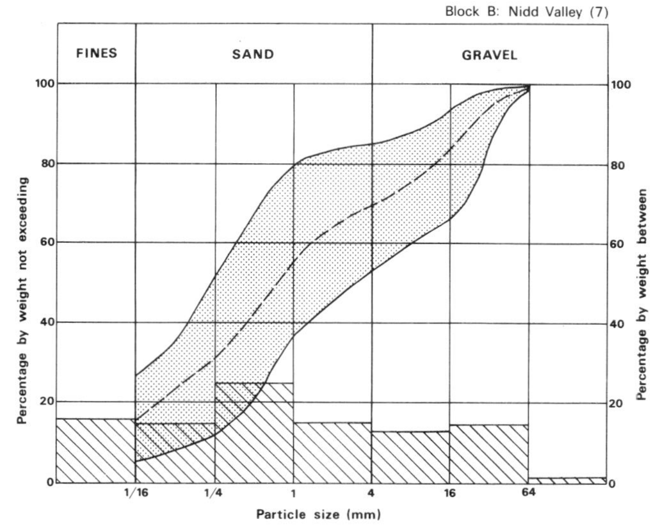

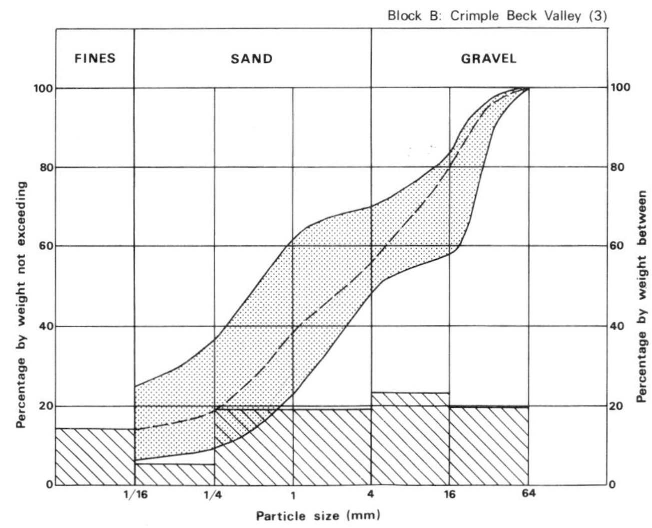

Figure 7 Mean particle-size distribution for the assessed thickness of mineral in Block B and for the mineral subdivisions within it.

The numbers in parentheses indicate the numbers of IMAU boreholes used to construct the diagrams. The shaded areas indicate the envelope within which the respective deposits fall. The stippled columns indicate the percentages that fall within the respective particle-size ranges.

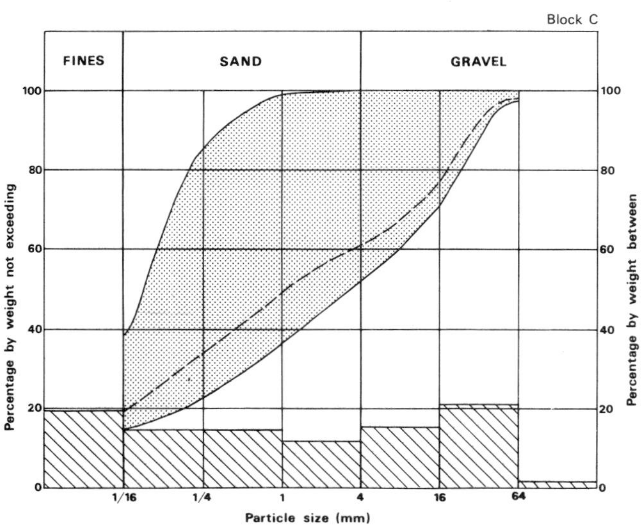

Figure 8 Mean particle-size distribution for the assessed thickness of mineral in Block C, based on four IMAU boreholes. The shaded areas indicate the envelope within which the deposits fall. The stippled columns indicate the percentages that fall within the respective particle-size ranges.

statistically assessed resource blocks the accuracy of the combined results at the 95 per cent probability level is 54 and 28 per cent respectively (that is, it is probable that 19 times out of 20 the true volumes of mineral present lie within the given limits). However, the true values are more likely to be nearer to the mean volumes estimated than to the limits. Moreover, it is probable that in each block approximately the same percentage limits would apply for the estimate of volume of a very much smaller parcel of ground (say 100 hectares) containing similar sand and gravel deposits, if the results from the same number of sample points (as provided by, say, ten boreholes) were used in the calculation. Thus, if closer limits are needed for the quotation of the reserves in part of a block, it can be expected that data from more than ten sample points will be required, even if the area is quite small.

It must again be emphasised that the assessed volume of sand and gravel resources has no simple relationship to the volume that could be extracted in practice, since no allowance has been made in the calculations for any restraints that may exist on the use of land for mineral working.

NOTES ON THE RESOURCE BLOCKS

The mineral-bearing portion of the district has been divided into three resource blocks (see the resource map and the statistical summary of the results of the assessment given in Table 6). Block A contains the Fluvio-glacial and Older River Sand and Gravel with associated laminated clays. In much of the block these deposits are overlain by till containing bodies of Glacial Sand and

Gravel or by late-Glacial clays, silts and sands of the Glacial Lake Deposits. However, along the northern margin of the block, except near the gravel workings north of Knaresborough, the till and lacustrine deposits directly overlie bedrock. Block B contains gravels that were laid down by Crimple Beck and by the Nidd in its present valley, after it had changed course to flow through the gorge at Knaresborough. Block C contains the remaining deposits of Glacial Sand and Gravel which are intercalated with till that directly overlies bedrock in the south-eastern and north-western corners of the district.

Block A (Table 3 and Figures 5 and 6)
This block, which has a total area of 20.3 km^2, contains the active sand and gravel workings of the district, as well as most of the deposits with the best potential for working in the future. The assessment is based on the records of 17 IMAU boreholes, of which 10 proved mineral. The block contains sand and gravel belonging to three distinct deposits: Fluvio-glacial and Older River Sand and Gravel; Glacial Sand and Gravel; and Glacial Lake Deposits, with associated Blown Sand. Since these deposits overlie one another in places, the mineral in each is assessed separately; confidence limits are quoted only for the Fluvio-glacial and Older River Sand and Gravel, since the borehole data are inadequate to allow more than an inferred assessment to be made of the mineral in the other two deposits.

The *Fluvio-glacial and Older River Sand and Gravel* that is currently being worked on the north-eastern outskirts of Knaresborough is there relatively thick, mostly between about 10 m and 23 m, with overburden usually between 2 m and 5 m thick, and with waste partings either thin or absent. However, as these channel deposits of the former course of the Nidd are followed south-eastwards, overburden thickens and the ratio of waste to mineral increases. Moreover, the overall grading of the mineral deteriorates owing to the presence of layers of clayey sand which often overlies the better-quality gravel, as, for instance, in boreholes NE 22 and NE 32.

The Fluvio-glacial and Older River Sand and Gravel and associated clays are mineral-bearing over an area of about 8.2 km^2, excluding about 1.4 km^2 of former and current gravel workings. Their mineral thickness proved in IMAU boreholes ranges from 1.4 m to more than 8.4 m, with a mean of 5.4 m. However, records of well borings (NE 5 and 6) indicate a mineral thickness of 22.5 m north-east of Knaresborough and 25.0 m at Castle Farm [382 583]. On the basis of the data from the IMAU boreholes alone, the volume of mineral belonging to the Fluvio-glacial and Older River Sand and Gravel present in the block is estimated to be 42.1 $\pm$ 23 million m^3. However, if the data from well records NE 5, 6, 7 and 9 are taken into consideration, and no adverse weighting applied, the estimated total volume is more than doubled, to 88.3 $\pm$ 55 million m^3. The considerable thicknesses of overburden (which ranges from 2.2 m to 13.7 m, with a mean of 6.7 m) and of waste (ranging from nil to 10.4 m, with a mean of 2.4 m), particularly in the eastern part of the block, reduces the potential value of these resources. Nevertheless, in those places where the Fluvio-glacial and Older River Sand and Gravel are overlain by younger mineral-bearing deposits, especially the gravels of the Glacial Sand and Gravel, as for example in borehole NE 22, their potential value and workability is much enhanced.

The Fluvio-glacial and Older River Sand and Gravel and their associated laminated clays are concealed for the most part by a cover of Till which contains, and is overlain by, numerous bodies of *Glacial Sand and Gravel*, particularly north of Goldsborough. Owing to the uncertainties resulting from their lenticular form and from the scarcity of borehole data, no estimate is given for their mean thickness in this block. Instead, individual bodies or groups of bodies have been assessed separately.

The group of bodies of Glacial Sand and Gravel that lie north, east and south of Hopewell House (i.e. at Sand Hills [366 592], Near and Far Andrew Hill [374 592], and Sixteen Acre Hill [377 587]) are thought, on the basis of their topographic expression, to contain significant thicknesses, perhaps up to 8 m in places, of mineral consisting mainly, on the evidence of hand augering, of more or less clayey sand, although an IMAU borehole (NE 22) south of Hopewell House proved 6.3 m of 'clayey' gravel. The above-mentioned deposits extend to about 45 hectares and may contain a total of about 1.8 million m^3 of mineral.

Two larger patches of Glacial Sand and Gravel were penetrated by IMAU boreholes. Of these, the deposit that underlies Spring Bank Farm [397 585] contains a potentially valuable gravel resource. Borehole NE 24 proved 22.4 m of 'clayey' gravel, of which about 12 m is above the water table. Overburden is only 0.4 m thick and there are no waste partings. The deposit probably extends to about 40 hectares and is estimated to contain about 6 million m^3 of mineral. There is a possibility that further gravel deposits belonging to the Fluvio-glacial and Older River Sand and Gravel might underlie this deposit: borehole NE 24 did not reach the bedrock. The other large patch of Glacial Sand and Gravel is at Ribston Park [39 54] and extends to about 120 hectares; IMAU boreholes NE 36 and SE 19 proved 6.4 m and 5.5+m respectively of mineral, comprising 'clayey' or 'very clayey' sand or pebbly sand with a total of 2.0 m of sandy clay waste partings in each case. The deposit is estimated to contain about 5 million m^3 of mineral.

The remaining patches of Glacial Sand and Gravel in the block are fairly numerous but small; they occur mostly in the area between Hall Farm, Flaxby and Goldsborough, and have a total area of about 50 hectares. There is virtually no evidence available as to their thickness, but, assuming a mean thickness of mineral of 2 m, they would contain a total of only about 1 million m^3 of mineral, much of which is believed, on the basis of hand-augering, to be 'clayey' sand.

The mineral in the *Glacial Lake Deposits* and associated *Blown Sand* west [38 58 and 38 59] of Coneythorpe and Flaxby extends to about 75 hectares; it was proved by only one IMAU borehole (NE 23), which penetrated 4.0 m of 'clayey' sand. The deposit is estimated to contain between 2 and 3 million m^3 of mineral.

The total volume of mineral present in the block is estimated to be between 34 and 156 million m^3; the wide range between these figures reflects the lenticular shape of many of the bodies of mineral; there are further uncertainties, not reflected in these figures, related to the extent of the mineral-bearing area. Moreover, owing to the considerable thickness of overburden in much of this area, it is probable that only a small volume, amounting perhaps to no more than about 15 million m^3, out of the total volume of mineral present could be commercially attractive in the foreseeable future.

Block B (Table 4 and Figures 5 and 7)
The total area of the block, which contains the fluvial and fluvio-glacial gravels of the valleys of the Nidd and

Crimple Beck, amounts to 9.5 km^2, of which 6.0 km^2 is considered to be mineral-bearing. Its assessment is based on the records of 13 IMAU boreholes, of which 10 proved mineral. The mineral occurs in five distinct deposits, namely (in order, from oldest to youngest): Fluvio-glacial and Older River Sand and Gravel (in boreholes SE 17, SE 22 and SE 25: the sand and gravel belonging to this deposit in borehole SE 20 has too much overburden to be classed as mineral); Glacial Sand and Gravel (borehole SE 16); Fluvio-glacial Terrace Deposits (boreholes NE 23 and NE 29), which have been worked in the past around Haughs Farm [366 563]; River Terrace Deposits (borehole SE 22); and Alluvium (boreholes NE 30, SE 21 and SE 28).

The borehole data are insufficient to allow the calculation of reliable figures for the mean thickness of the mineral in the individual deposits, although calculated mean thicknesses have been given in Table 6. However, the overall mean thickness of mineral present in the block is calculated to be 3.1 m $\pm$ 28 per cent. Taken in conjunction with the figure of 6.0 km^2 for the extent of the mineral-bearing area, this mean thickness indicates a total volume of 18.6 $\pm$ 5 million m^3 for the mineral estimated to be present in the block. However, the actual volume of mineral present may well be lower than this, since there is considerable uncertainty as to the extent of the mineral-bearing ground in the valley of Crimple Beck.

Probably the only deposits that might be commercially attractive in the foreseeable future are the terrace deposits in the lower reaches of Crimple Beck north-west and north-north-east of North Deighton [390 516]; in the Nidd valley the mineral-bearing ground is restricted to rather narrow strips on either side of the river, except perhaps north of Little Ribston [386 534], where the depth, thickness and grading of the mineral have not yet been proved by boreholes.

Block C (Table 5 and Figures 5 and 8)

The remainder of the potentially mineral-bearing ground in the district is designated as Block C; it has a total area of 18.9 km^2 and is made up of two widely separated sub-blocks: the larger sub-block, C^1, lying north of Harrogate, covers 13.4 km^2, and the smaller, C^2, around Kirk Deighton in the south-eastern corner of the district, extends to 4.7 km^2.

The borehole data on the sand and gravel present in Block C are limited, since only three of the eight IMAU boreholes in the block proved mineral; in each of these three the mineral proved belonged to the Glacial Sand and Gravel. This deposit forms irregularly shaped bodies within the till that masks the bedrocks over much of the block. Because of their unpredictable shape, the sub-surface extent of these bodies is very uncertain and may often bear little relationship to their mapped surface outcrop. On the basis of the presently available information, the only body of mineral that is known to be of significant extent and thickness in the Glacial Sand and Gravel of this block is that which lies north of Scotton [325 593]. The areal extent of this body, whose eastern part has been worked in the past, is not known with any accuracy, but may amount to about 30 hectares. Although borehole NW 23, west-north-west of Scotton, proved 3.7 m of 'very clayey' gravel, the environs of this borehole have not been depicted as mineral-bearing, since the extent of the mineral-bearing ground is unknown (mineral is absent in the adjacent borehole NW 24), and the thickness of overburden (9.7 m) would anyway probably render it a commercially unattractive deposit.

Unassessed patches of Alluvium and Fluvio-glacial Terrace Deposits in the Nidd valley in Sub-block C^1 may also contain mineral, but, apart from areas amounting to about 15 hectares on either bank of the Nidd west of Viaduct Wood [305 584] and an area of about 13 hectares on its right bank opposite Conyngham Hall [342 574], they are of insignificant extent, and the volume of any sand and gravel they may contain must be negligible. The patch of Fluvio-glacial Terrace Deposits that fills the Kirk Deighton Channel [395 503] extends to about 10 hectares and may perhaps contain a small deposit of mineral, but the volume present is unlikely to amount to much more than about 0.1 million m^3. A similar volume of mineral might be available elsewhere in Sub-block C^2 in the several small patches of Glacial Sand and Gravel that crop out around North Deighton.

APPENDIX A

FIELD AND LABORATORY PROCEDURES

Trial and error during initial studies of the complex and variable glacial deposits of East Anglia and Essex showed that an absolute minimum of five sample points evenly distributed across the sand and gravel are needed to provide a worthwhile statistical assessment, but that, where possible, there should be not less than ten. Sample points are any points for which adequate information exists about the nature and thickness of the deposit and may include boreholes other than those drilled during the survey and exposures. In particular, the cooperation of sand and gravel operators ensures that boreholes are not drilled where reliable information is already available; although this may be used in the calculations, it is held confidentially by the Institute and cannot be disclosed.

The mineral shown on each 1 : 25 000 sheet is divided into resource blocks. The arbitrary size selected, $10\,km^2$, is a compromise to meet the aims of the survey by providing sufficient sample points in each block. As far as possible the block boundaries are determined by geological boundaries so that, for example, glacial and river terrace gravels are separated. Otherwise division is by arbitrary lines, which may bear no relationship to the geology. The blocks are drawn provisionally before drilling begins.

A reconnaissance of the ground is carried out to record any exposures and inquiries are made to ascertain what borehole information is available. Borehole sites are then selected to provide an even pattern of sample points at a density of approximately one per square kilometre. However, because broad trends are independently overlain by smaller-scale characteristically random variations, it is unnecessary to adhere to a square grid pattern. Thus such factors as ease of access and the need to minimise disturbance to land and the public are taken into account in siting the holes; at the same time it is necessary to guard against the possibility that ease of access (that is, the positions of roads and farms) may reflect particular geological conditions, which may bias the drilling results.

The drilling machine employed should be capable of providing a continuous sample representative of all unconsolidated deposits, so that the in-situ grading can be determined, if necessary, to a depth of 30 m (100 ft) at a diameter of about 200 mm (8 in), beneath different types of overburden. It should be reliable, quiet, mobile and relatively small (so that it can be moved to sites of difficult access). Shell and auger rigs have proved to be almost ideal.

The rigs are modified to enable deposits above the water table to be drilled 'dry', instead of with water added to facilitate the drilling, to minimise the amount of material drawn in from outside the limits of the hole. The samples thus obtained are representative of the in-situ grading, and satisfy one of the most important aims of the survey. Below the water table the rigs are used conventionally, although this may result in the loss of some of the fines fraction and the pumping action of the bailer tends to draw unwanted material into the hole from the sides or the bottom.

A continuous series of bulk samples is taken throughout the sand and gravel. Ideally samples are composed exclusively of the whole of the material encountered in the borehole between stated depths. However, care is taken to discard, as far as possible, material which has caved or has been pumped from the bottom of the hole. A new sample is commenced whenever there is an appreciable lithological change within the sand and gravel, or at every 1 m (3.3 ft) depth. The samples, each weighing between 25 and 45 kg (55 and 100 lb), are despatched in heavy duty polythene bags to a laboratory for grading. The grading procedure is based on B.S. 1377 (British Standards Institution, 1967). Random checks on the accuracy of the grading are made in the Institute's laboratories.

All data, including mean grading analysis figures calculated for the total thickness of the mineral, are entered on standard record sheets, abbreviated copies of which are reproduced in Appendix F.

Detailed records may be consulted at the appropriate offices of the Institute, upon application to the Head, Industrial Minerals Assessment Unit.

APPENDIX B

STATISTICAL PROCEDURE

Statistical assessment

1 A statistical assessment is made of an area of mineral greater than $2\,km^2$, if there is a minimum of five evenly spaced boreholes in the resource block (for smaller areas see paragraph 12 below).

2 The simple methods used in the calculations are consistent with the amount of data provided by the survey. Conventional symmetrical confidence limits are calculated for the 95 per cent probability level, that is, there is a 5 per cent or one in twenty chance of a result falling outside the stated limits.

3 The volume estimate (V) for the mineral in a given block is the product of the two variables, the sampled areas (A) and the mean thickness ($\bar{l}_m$) calculated from the individual thicknesses at the sample points. The standard deviations for these variables are related such that

$$S_V = \sqrt{(S_A{}^2 + S_{\bar{l}_m}{}^2)} \quad . \qquad [1]$$

4 The above relationship may be transposed such that

$$S_V = S_{\bar{l}_m}\sqrt{(1 + S_A{}^2/S_{\bar{l}_m}{}^2)} \quad . \qquad [2]$$

From this it can be seen that as $S_A{}^2/S_{\bar{l}_m}{}^2$ tends to 0, S_V tends to $S_{\bar{l}_m}$.

If, therefore, the standard deviation for area is small with respect to that for mean thickness, the standard deviation for volume approximates to that for mean thickness.

5 Given that the number of approximately evenly spaced sample points in the sampled area is n, with mineral thickness measurements $l_{m_1}, l_{m_2}, \ldots, l_{m_n}$ then the best estimate of mean thickness, $\bar{l}_m$, is given by

$$\Sigma\,(l_{m_1} + l_{m_2} \ldots l_{m_n})/n.$$

For groups of closely spaced boreholes a discretionary weighting factor may be applied to avoid bias (see note on weighting below). The standard deviation for mean thickness $S_{\bar{l}}$, expressed as a proportion of the mean thickness, is given by

$$S_l = (1/\bar{l}_m)\sqrt{[\Sigma(l_m - \bar{l}_m)^2/(n-1)]}$$

where l_m is any value in the series l_{m_1} to l_{m_n}.

6 The sampled area in each resource block is coloured pink on the map. Wherever possible, calculations relate to the mineral within mapped geological boundaries (which may not necessarily correspond to the limits of deposit). Where the area is not defined by a mapped boundary, that is, where the boundary is inferred, a distinctive symbol is used. Experience suggests that the errors in determining area are usually small relative to those in thickness. The relationship $S_A/S\bar{l}_m \leqslant \frac{1}{3}$ is assumed in all cases. It follows from equation [2] that

$$S_{\bar{l}_m} \leqslant S_V \leqslant 1.05\, S_{\bar{l}_m}. \qquad [3]$$

7 The limits on the estimate of mean thickness of mineral, $L_{\bar{l}_m}$, may be expressed in absolute units $\pm(t/\sqrt{n}) \times S_{l_m}$ or as a percentage $\pm(t/\sqrt{n})t\,S_{\bar{l}_m} \times (100/\bar{l}_m)$ per cent, where t is Student's t at the 95 per cent probability level for $(n-1)$ degrees of freedom, evaluated by reference to statistical tables. (In applying Student's t it is assumed that the measurements are distributed normally).

Block calculation 1 : 25 000 block: Fictitious

Area
Block: 11.08 km^2
Mineral: 8.32 km^2

Mean thickness
Overburden: 2.5 m
Mineral: 6.5 m

Volume
Overburden: 21 million m^3
Mineral: 54 million m^3

Confidence limits of the estimate of mineral volume at the 95 per cent probability level: ± 20 per cent
That is, the volume of mineral (with 95 per cent probability): 54 ± 11 million m^3

Thickness estimate measurements in metres l_o = overburden thickness l_m = mineral thickness

Sample point	Weighting w	Overburden l_o	Overburden wl_o	Mineral l_m	Mineral wl_m	Remarks
SE 14	1	1.5	1.5	9.4	9.4	IMAU boreholes
SE 18	1	3.3	3.3	5.8	5.8	
SE 20	1	nil	–	6.9	6.9	
SE 22	1	0.7	0.7	6.4	6.4	
SE 23	1	6.2	6.2	4.1	4.1	
SE 24	1	4.3	4.3	6.4	6.4	
SE 17	$\frac{1}{2}$	1.2	1.6	9.8	7.2	Hydrogeology Unit record
123/45	$\frac{1}{2}$	2.0		4.6		
1	$\frac{1}{4}$	2.7	2.6	7.3	5.8	Close group of four boreholes (commercial)
2	$\frac{1}{4}$	4.5		3.2		
3	$\frac{1}{4}$	0.4		6.8		
4	$\frac{1}{4}$	2.8		5.9		
Totals	$\Sigma w = 8$		$\Sigma wl_o = 20.2$		$\Sigma wl_m = 52.0$	
Means			$\overline{wl_o} = 2.5$		$\overline{wl_m} = 6.5$	

Calculation of confidence limits

wl_m	$\lvert(wl_m - \overline{wl_m})\rvert$	$(wl_m - \overline{wl_m})^2$
9.4	2.9	8.41
5.8	0.7	0.49
6.9	0.4	0.16
6.4	0.1	0.01
4.1	2.4	5.76
6.4	0.1	0.01
7.2	0.7	0.49
5.8	0.7	0.49

$\Sigma(wl_m - \overline{wl_m})^2 = 15.82$
$n = 8$
t = 2.365

L_V is calculated as

$$1.05\,(t/\overline{wl_m})\sqrt{[\Sigma(wl_m - \overline{wl_m})^2/n(n-1)]} \times 100$$
$$= 1.05 \times (2.365/6.5)\sqrt{[15.82/(8 \times 7)]} \times 100$$
$$= 20.3$$
$$\simeq 20 \text{ per cent}$$

Figure 9 Example of resource block assessment: calculation and results.

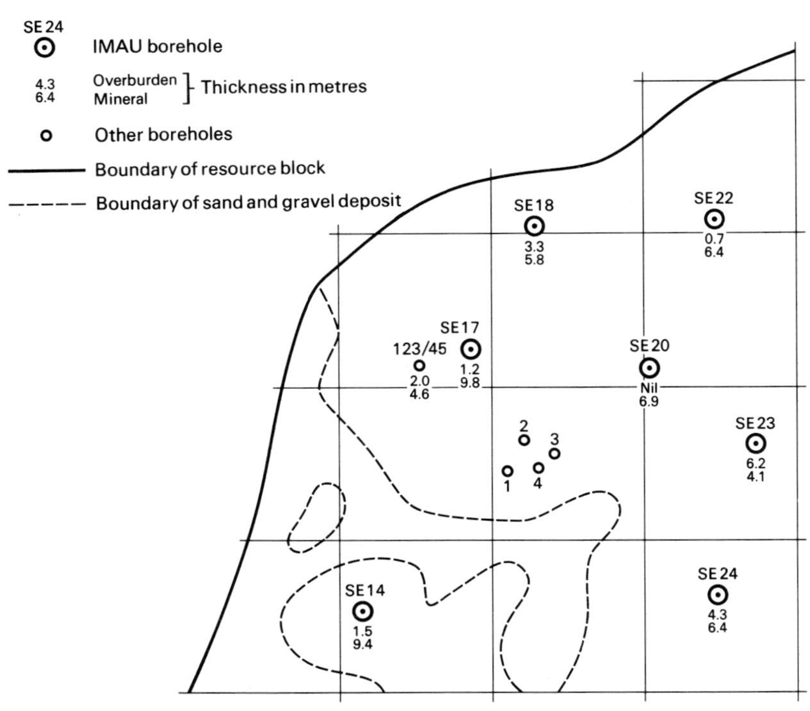

Figure 10 Example of resource block assessment: map of a fictitious block.

8 Values of t at the 95 per cent probability level for values of n up to 20 are as follows:

n	t	n	t
1	infinity	11	2.228
2	12.706	12	2.201
3	4.303	13	2.179
4	3.182	14	2.160
5	2.776	15	2.145
6	2.571	16	2.131
7	2.447	17	2.120
8	2.365	18	2.110
9	2.306	19	2.101
10	2.262	20	2.093

(from Table 12, Biometrika Tables for Statisticians, Volume 1, Second Edition, Cambridge University Press, 1962). When n is greater than 20, 1.96 is used (the value of t when n is infinity).

9 In calculating confidence limits for volume, L_V, the following inequality corresponding to equation [3] is applied: $L_{\bar{l}_m} \leqslant L_V \leqslant 1.05\, L_{\bar{l}_m}$

10 In summary, for values of n between 5 and 20, L_V is calculated as

$$[(1.05 \times \mathrm{t})/\bar{l}_{\mathrm{m}}] \times [\sqrt{\Sigma(l_{\mathrm{m}} - \bar{l}_{\mathrm{m}})^2/n(n-1)}] \times 100 \text{ per cent,}$$

and when n is greater than 20, as

$$[(1.05 \times 1.96)/\bar{l}_{\mathrm{m}}] \times [\sqrt{\Sigma(l_{\mathrm{m}} - \bar{l}_{\mathrm{m}})^2/n(n-1)}] \times 100 \text{ per cent.}$$

(weighting factors may be included: see paragraph 15).

11 The application of this procedure to a fictitious area is illustrated in Figures 9 and 10.

Inferred assessment

12 If the sampled area of mineral in a resource block is between 0.25 km^2 and 2 km^2 an assessment is inferred, based on geological and topographical information usually supported by the data from one or two boreholes. The volume of mineral is calculated as the product of the area, measured from field data, and the estimated thickness. Confidence limits are not calculated.

13 In some cases a resource block may include an area left uncoloured on the map, within which mineral (as defined) is interpreted to be generally absent. If there is reason to believe that some mineral may be present, an inferred assessment may be made.

14 No assessment is attempted for an isolated area of mineral less than 0.25 km^2.

15 *Note on weighting* The thickness of a deposit at any point may be governed solely by the position of the point in relation to a broad trend. However, most sand and gravel deposits also exhibit a random pattern of local, and sometimes considerable, variation in thickness. Thus the distribution of sample points need be only approximately regular and in estimating the mean thickness only simple weighting is necessary. In practice, equal weighting can often be applied to thicknesses at all sample points. If, however, there is a distinctly unequal distribution of points, bias is avoided by dividing the sampled area into broad zones, to each of which a value roughly proportional to its area is assigned. This value is then shared between the data points within the zone as the weighting factor.

APPENDIX C

CLASSIFICATION AND DESCRIPTION OF SAND AND GRAVEL

For the purposes of assessing resources of sand and gravel a classification should take account of economically important characteristics of the deposit, in particular the absolute content of fines and the ratio of sand to gravel.

The terminology commonly used by geologists when describing sedimentary rocks (Wentworth, 1922) is not entirely satisfactory for this purpose. For example, Wentworth proposed that a deposit should be described as a 'gravelly sand' when it contains more sand than gravel and there is at least 10 per cent of gravel, provided that there is less than 10 per cent of material finer than sand (less than $\frac{1}{16}$ mm) and coarser than pebbles (more than 64 mm in diameter). Because deposits containing more than 10 per cent fines are not embraced by this system a modified binary classification based on Willman (1942) has been adopted.

When the fines content exceeds 40 per cent the material is not considered to be potentially workable and falls outside the definition of mineral. Deposits which contain 40 per cent fines or less are classified primarily on the ratio of sand to gravel but qualified in the light of the fines content, as follows: less than 10 per cent fines—no qualification; 10 per cent or more but less than 20 per cent fines—'clayey'; 20 to 40 per cent fines—'very clayey'.

The term 'clay' (as written, with single quote marks) is used to describe all material passing $\frac{1}{16}$ mm. Thus it has no mineralogical significance and includes particles falling within the size range of silt. The normal meaning applies to the term clay where it does not appear in single quotation marks.

The ratio of sand to gravel defines the boundaries between sand, pebbly sand, sandy gravel and gravel (at 19 : 1, 3 : 1 and 1 : 1).

Thus it is possible to classify the mineral into one of twelve descriptive categories (see Figure 11). The procedure is as follows:

1 Classify according to ratio of sand to gravel.
2 Describe fines.

For example, a deposit grading 11 per cent gravel, 70 per cent sand and 19 per cent fines is classified as 'clayey' pebbly sand. This short description is included in the borehole log (see Note 10, Appendix D).

Many differing proposals exist for the classification of the grain size of sediments (Atterberg, 1905; Udden, 1914; Wentworth, 1922; Wentworth, 1935; Allen, 1936; Twenhofel, 1937; Lane and others, 1947). As Archer (1970a, b) has emphasised, there is a pressing need for a simple metric scale acceptable to both scientific and engineering interests, for which the class limit sizes correspond closely with certain marked changes in the natural properties of mineral particles. For example, there is an important change in the degree of cohesion between particles at about the $\frac{1}{16}$-mm size, which approximates to the generally accepted boundary between silt and sand. These and other requirements are met by a system based on Udden's geometric scale and a simplified form of Wentworth's terminology (Table 7), which is used in this Report.

The fairly wide intervals in the scale are consistent with the general level of accuracy of the qualitative assessments of the resource blocks. Three sizes of sand are recognised, fine ($+\frac{1}{16}-\frac{1}{4}$ mm), medium ($+\frac{1}{4}-1$ mm) and coarse ($+1-4$ mm). The boundary at 16 mm distinguishes a range of finer gravel ($+4-16$ mm), often characterised by abundance of worn tough pebbles of vein quartz, from larger pebbles often of notably different materials. The boundary at 64 mm distinguishes pebbles from cobbles. The term 'gravel' is used loosely to denote both pebble-sized and cobble-sized material.

The size distribution of borehole samples is determined by sieve analysis, which is presented by the laboratory as logarithmic cumulative curves (see, for example, British

Standard 1377: 1967). In this report the grading is tabulated on the borehole record sheets (Appendix F), the intercepts corresponding with the simple geometric scale $\frac{1}{16}$ mm, $\frac{1}{4}$ mm, 1 mm, 4 mm, 16 mm and so on as required. Original sample grading curves are available for reference at the appropriate office of the Institute.

Each bulk sample is described, subjectively, by a geologist at the borehole site. Being based on visual examination, the description of the grading is inexact, the accuracy depending on the experience of the observer. The descriptions recorded are modified, as necessary, when the laboratory results become available.

The relative proportions of the rock types present in the gravel fraction are indicated by the use of the words 'and' or 'with'. For example, 'flint and quartz' indicates roughly equal proportions with neither constituent accounting for less than about 25 per cent of the whole; 'flint with quartz' indicates that flint is dominant and quartz, the principal accessory rock type, comprises 5 to 25 per cent of the whole. Where the accessory material accounts for less than 5 per cent of the whole, but is still readily apparent, the phrase 'with some' has been used. Rare constituents are referred to as 'trace'.

The terms used in the field to describe the degree of rounding of particles, which is concerned with the sharpness of the edges and corners of a clastic fragment and not the shape (after Pettijohn, 1957), are as follows.

Angular: showing little or no evidence of wear; sharp edges and corners.

Subangular: showing definite effects of wear. Fragments still have their original form but edges and corners begin to be rounded off.

Subrounded: showing considerable wear. The edges and corners are rounded off to smooth curves. Original grain shape is still distinct.

Rounded: original faces almost completely destroyed, but some comparatively flat surfaces may still remain. All original edges and corners have been smoothed off to rather broad curves. Original shape is still apparent.

Well-rounded: no original faces, edges or corners left. The entire surface consists of broad curves; flat areas are absent. The original shape is suggested by the present form of the grain.

Table 7 Classification of gravel, sand and fines

Size limits	Grain size description	Qualification	Primary classification
	Cobble		
64 mm			
		Coarse	Gravel
16 mm	Pebble		
		Fine	
4 mm			
		Coarse	
1 mm			
	Sand	Medium	Sand
$\frac{1}{4}$ mm			
		Fine	
$\frac{1}{16}$ mm			
	Fines (silt and clay)		Fines

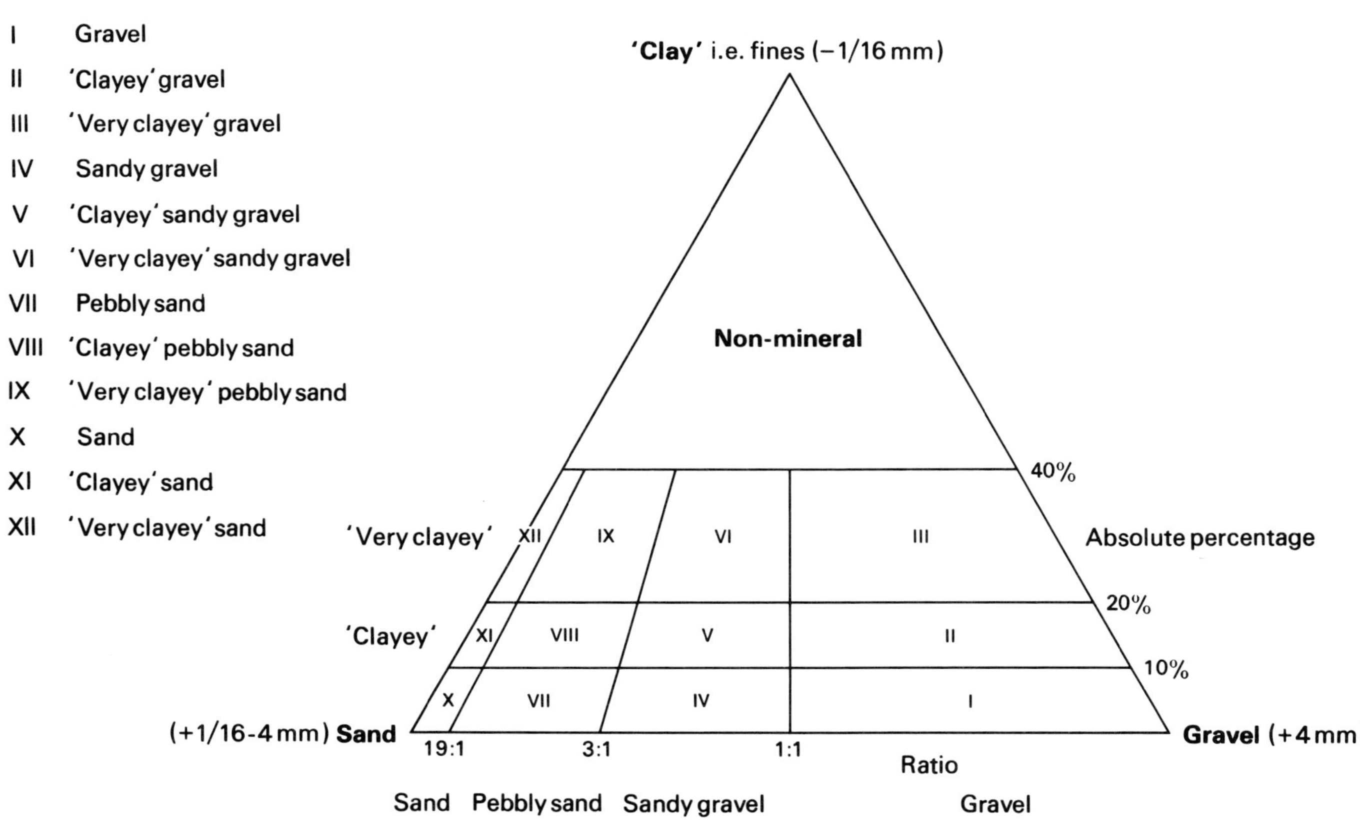

Figure 11 Diagram showing the descriptive categories used in the classification of sand and gravel.

APPENDIX D

EXPLANATION OF THE BOREHOLE RECORDS

Annotated example

SE 35 NE 32[1] **3972 5656**[2] **Goldsborough Moor**[3] **Block A**

Surface level (+30.5 m) +100 ft[4]
Water struck at (+24.0 m) +79 ft, artesian[5]
203 mm[6]
April 1979

Overburden[7] 0.5 m
Mineral 1.0 m
Waste 13.7 m
Mineral 4.6 m
Bedrock 1.4+ m[8]

LOG

Geological classification[9]	Lithology[10]	Thickness m	Depth m
	Soil, dusky brown, silty clay	0.2	0.2
Silt and Clay glacial lake deposits	Clay, greyish brown, silty	0.3	0.5
Glacial Sand and Gravel	**a** 'Clayey' sandy gravel Gravel: fine and coarse, subrounded to rounded quartzitic sandstone, with black subangular limestone Sand: medium, fine and coarse; subangular quartz, quartzitic sandstone and black limestone	1.0	1.5
Till	Clay, sandy, with pebbles	0.2	1.7
Fluvio-glacial and Older River Sand and Gravel and associated clay	Clay, brown and greyish brown, silty, laminated, with a layer of 'very clayey' sand (**b**) from 8.2 to 10.0 m and of 'very clayey' pebbly sand (**c**) from 12.8 to 14.5 m	13.5	15.2
	d Sandy gravel Gravel: fine and coarse, rounded and well rounded sandstone, siltstone and dolomitic limestone, with some black limestone Sand: medium, coarse and fine, mainly subangular quartz with subrounded limestone and subangular dolomitic limestone	4.6	19.8
Upper Marl?	Sand, red-brown, clayey, fine-grained	0.9	20.7
Upper Marl	Marl, dark reddish brown, with yellowish green mottles and laminate	0.5+	21.2

GRADING[11]

	Mean for deposit *percentages*			Depth below surface (m)	*percentages*						
	Fines	Sand	Gravel		Fines	Sand			Gravel		
					$-\frac{1}{16}$ mm	$+\frac{1}{16}-\frac{1}{4}$ mm	$+\frac{1}{4}-1$ mm	+1–4 mm	+4–16 mm	+16–64 mm	+64 mm
a	20	46	34	0.5–1.5	20	17	19	10	24	10	0
b*	33	67	0	8.2–9.0	32	56	12	0	0	0	0
				9.0–10.0	35	49	15	1	0	0	0
				Mean	33	52	14	1	0	0	0
c*	27	62	11	12.8–13.5	21	18	27	12	10	12	0
				13.5–14.5	31	49	15	2	3	0	0
				Mean	27	36	20	6	6	5	0
d	6	64	30	15.2–15.8	13	21	63	3	0	0	0
				15.8–16.8	7	16	38	5	16	16	3
				16.8–17.8	4	10	34	13	21	18	0
				17.8–18.8	4	6	36	15	21	19	0
				18.8–19.8	6	10	48	13	13	10	0
				Mean	6	12	42	10	16	13	1
a+b+c+d	17	62	21	Mean	17	25	30	7	12	9	0

* Non-mineral: excessive overburden.

COMPOSITION[12]

	Depth below surface (m)	*Percentages by weight in gravel fraction*					
		Limestone	Dolomitic limestone	Sandstone	Quartz	Siltstone	Ironstone
d	16.8–17.8	3	11	67	4	14	1

The numbered paragraphs below correspond with the annotations given on the specimen record.

1 Borehole Registration Number
Each Industrial Minerals Assessment Unit (IMAU) borehole is identified by a Registration Number. This consists of two statements.

1 The number of the 1:25 000 sheet on which the borehole lies, for example SE 35.

2 The quarter of the 1:25 000 sheet on which the borehole lies and the number of the borehole in a series for that quarter, for example NE 32.

Thus the full Registration Number is SE 35 NE 32. This is abbreviated to NE 32 in the text.

2 The National Grid reference
All National Grid references in this publication lie within the 100-km square SE unless otherwise stated. Grid references are given to eight figures, accurate to within 10 m for borehole locations. (In the text, six- or four-figure grid references are used for more extensive locations, for example, for farms or hills).

3 Location
The position of the borehole is generally referred to the nearest named locality on the 1:25 000 base map and the resource block in which it lies is stated.

4 Surface level
The surface level at the borehole site is given in metres and feet above Ordnance Datum. All measurements were made in feet; approximate conversions to metres are given in brackets.

5 Groundwater conditions
If groundwater was present the level at which it was encountered is normally given (in metres and feet above OD).

6 Type of drill and date of drilling
All boreholes were drilled by a shell-and-auger rig using either 152-mm or 203-mm casing. The month and year of completion of the hole are stated.

7 Overburden, mineral, waste and bedrock
Mineral is sand and gravel that falls within the definition of potentially workable material (see p. 1). Bedrock refers to the usually consolidated rocks listed in Table 1 as belonging to the Solid formations of Carboniferous to Triassic age, in which mineral would not normally be expected to be present. In fact, some of the strata belonging to the Sherwood Sandstone Group are poorly consolidated when weathered and consist of 'clayey' sand, but no attempt has been made to assess resources of this material and, for the purposes of this survey, it has not been classified as mineral. Waste is any material other than bedrock or mineral; where waste occurs between the surface and the top of a deposit of mineral it is classified as overburden.

8 The plus sign (+) indicates that the base of the deposit was not reached during drilling.

9 Geological classification
The geological classification (see Table 1) is given whenever possible.

10 Lithological description
When sand and gravel is recorded a general description based on the grading characteristics (for details see Appendix C) is followed by more detailed particulars. The description of other rocks is based on visual examination, in the field. Where more than one mineral deposit is recognised, each is designated by a letter, e.g. **a**, **b**, etc.

11 Grading data
A continuous series of bulk samples is taken throughout the thickness of sand and gravel. A new sample is commenced whenever there is an appreciable lithological change or at every 1 m of depth.

For each bulk sample the percentages of fines ($-\frac{1}{16}$ mm), fine sand ($+\frac{1}{16}$–$\frac{1}{4}$ mm), medium sand ($+\frac{1}{4}$–1 mm), coarse sand (+1–4 mm), fine gravel (+4–16 mm) and coarse gravel (+16 mm) are stated. The mean grading of groups of samples making up an identified mineral horizon are also given in detail and, to the left, in summary. Where more than one horizon is recognised the mean grading for the whole of the mineral in the borehole is also given. Where necessary in calculating the mean grading, data for individual samples are weighted by the thickness represented.

Fully representative sampling of sand and gravel is difficult to achieve, particularly where groundwater levels are high. Comparison between boreholes and adjacent exposures suggests that in borehole samples the proportion of sand may be higher and the proportion of fines and coarse gravel may be lower.

12 Composition
Details of the composition of the gravel fraction of selected samples or grouped samples may be given. Where appropriate the calculated weighted mean composition of grouped samples may be indicated.

APPENDIX E

LIST OF BOREHOLES USED IN THE ASSESSMENT OF RESOURCES

Borehole number	Grid references	Resource block
1 MINERAL ASSESSMENT BOREHOLES		
NW 22	3004 5985	C[1]
23	3153 5979	C[1]
24	3232 5979	C[1]
25	3285 5987	C[1]
26	3489 5833	C[1]
NE 17	3700 5977	A
18	3758 5945	A
19	3851 5951	A
20	3975 5932	A
21	3683 5813	A
22	3723 5848	A
23	3898 5889	A
24	3966 5855	A
25	3689 5748	A
26	3794 5770	A
27	3851 5718	A
28	3959 5758	A
29	3630 5604	B
30	3675 5661	B
31	3798 5670	A
32	3972 5656	A
33	3753 5519	B
34	3760 5582	B
35	3863 5576	A
36	3961 5510	A
SE 16	3737 5456	B
17	3822 5441	B
18	3864 5454	B
19	3938 5448	A
20	3872 5310	B
21	3960 5366	B
22	3990 5310	B
23	3870 5227	B
24	3960 5261	C[2]
25	3829 5181	B
26	3960 5168	C[2]
27	3983 5110	C[2]
28	3633 5156	B
2 OTHER BOREHOLES		
NE 5	3580 5868	A
6	3563 5918	A
7	3824 5835	A
9	3788 5930	A

APPENDIX F

INDUSTRIAL MINERALS ASSESSMENT UNIT BOREHOLE RECORDS

SE 35 NW 22 3004 5985 Pondhouse Farm, Nidd **Block C[1]**

Surface level (+72.6 m) +238 ft
Water struck at (+64.6 m) +212 ft
203 mm
March 1979

Waste 9.4 m
Bedrock 0.1 m+

LOG

Geological classification	Lithology	Thickness m	Depth m
	Sandy loam	0.1	0.1
Fluvio-glacial Terrace Deposits	Clay, sandy, gravelly, and sandy silt	5.0	5.1
Silt and Clay of the Glacial Lake Deposits	Clay, stiff, laminated, brown, with paler brown laminae of fine quartz sand and silt	2.5	7.6
Till	Clay, gravelly, with fine and coarse gravel of angular and subangular fragments of Magnesian Limestone	1.8	9.4
Lower Magnesian Limestone	Limestone, hard, light grey, dolomitic	0.1+	9.5

SE 35 NW 23 3153 5979 Brearton High Moor, Nidd **Block C[1]**

Surface level (+72.6 m) +238 ft
Water not encountered
203 mm
March 1979

Overburden 9.7 m
Mineral 3.7 m
Waste 7.8 m
Bedrock 0.1 m+

LOG

Geological classification	Lithology	Thickness m	Depth m
	Soil	0.6	0.6
Till	Clay, mottled brown, becoming dark greyish brown below 3.0 m, with some sand and coarse and fine gravel of sandstone and limestone; some cobbles	9.1	9.7
Glacial Sand and Gravel	'Clayey' gravel Gravel: coarse and fine, subangular to subrounded sandstone and subrounded to well rounded limestone Sand: coarse to fine, subangular to subrounded quartz with some sandstone and limestone lithic fragments in the coarser fraction	3.7	13.4
Laminated Clay closely associated with Glacial Deposits	Clay, brown, silty and sandy with beds of sand and silt and gravelly silty clay with pebbles of shale and micaceous sandstone	7.8	21.2
Namurian, undivided	Sandstone, yellowish grey, soft, flaggy, micaceous	0.1+	21.3

GRADING

Mean for deposit *percentages*			Depth below surface (m)	*percentages*						
Fines	Sand	Gravel		Fines	Sand			Gravel		
				$-\frac{1}{16}$	$+\frac{1}{16}-\frac{1}{4}$	$+\frac{1}{4}-1$	+1–4	+4–16	+16–64	+64
15	37	48	9.7–11.7	12	9	13	8	21	33	4
			11.7–13.4	19	6	15	25	20	16	0
			Mean	15	8	14	16	20	25	2

SE 35 NW 24 3232 5979 Old Hall Farm, Scotton **Block C[1]**

Surface level (+55.3 m) +181 ft
Water struck at (+45.5 m) +149 ft
203 mm
March 1979

Waste 8.6 m
Bedrock 2.1 m+

LOG

Geological classification	Lithology	Thickness m	Depth m
	Soil	0.3	0.3
Till	Clay, grey and brown, sandy and pebbly, with coarse and fine pebbles of dolomitic limestone, Carboniferous Limestone, sandstone and quartzite	0.8	1.1
Clay closely associated with Glacial Deposits Glacial Deposits	Clay, grey and brown, laminated with silt partings and a layer of fine and medium sand from 4.6 to 6.0 m	5.9	7.0
	Silt, grey, soft, sandy, micaceous, clayey	1.6	8.6
Namurian, undivided	Sandstone, yellowish brown, flaggy, fine- to medium-grained, micaceous, quartzitic	2.1+	10.7

GRADING

	Mean for deposit *percentages*			Depth below surface (m)	*percentages*						
	Fines	Sand	Gravel		Fines	Sand			Gravel		
					$-\frac{1}{16}$	$+\frac{1}{16}-\frac{1}{4}$	$+\frac{1}{4}-1$	+1–4	+4–16	+16–64	+64
*	25	75	0	4.6–6.0	25	60	15	0	0	0	0

* Non mineral: excessive overburden.

SE 35 NW 25 3285 5987 Lawrence House Farm, Scotton **Block C[1]**

Surface level (+52.7 m) +173 ft
Water struck at (+36.5 m) +120 ft
203 mm
March 1979

Overburden 2.4 m
Mineral 9.6 m
Waste 4.4 m
Bedrock 0.1 m+

LOG

Geological classification	Lithology	Thickness m	Depth m
Glacial Sand and Gravel	Soil passing down into brown sand with sandstone pebbles	1.4	1.4
	Silt, sandy, yellowish red	1.0	2.4
	'Clayey' gravel Gravel: coarse and fine, subangular sandstone, and subangular to subrounded Carboniferous and Magnesian Limestone, with some chert and quartz Sand: fine to coarse, quartz and lithics as in gravel fraction	9.6	12.0
Clay closely associated with Glacial deposits	Clay, stiff, olive-brown and dark grey mottled with shale fragments towards the base	4.4	16.4
Namurian, undivided	Shale, black	0.1+	16.5

GRADING

Mean for deposit percentages			Depth below surface (m)	percentages						
Fines	Sand	Gravel		Fines	Sand			Gravel		
				$-\frac{1}{16}$	$+\frac{1}{16}-\frac{1}{4}$	$+\frac{1}{4}-1$	+1–4	+4–16	+16–64	+64
17	36	47	2.4–4.4	16	11	11	10	22	31	0
			4.4–6.4	17	8	10	12	16	29	8
			6.4–7.4	18	10	12	15	24	20	1
			7.4–9.4	20	23	16	5	13	19	5
			9.4–12.0	16	15	13	11	20	25	1
			Mean	17	15	12	10	18	25	3

SE 35 NW 26 3489 5833 Park Corner Farm, Scriven **Block C[1]**

Surface level (+56.0 m) +184 ft
Water not encountered
203 mm
April 1979

Waste 4.3 m
Bedrock 2.7 m+

LOG

Geological classification	Lithology	Thickness m	Depth m
	Soil	0.3	0.3
Till	Clay, reddish brown, silty and sandy	4.0	4.3
Middle Marl	Marl, brownish red, silty	2.7+	7.0

SE 35 NE 17 3700 5977 Hogwood Lane, Ferrensby **Block A**

Surface level (+55.0 m) +180 ft
Water not encountered
203 mm
April 1979

Waste 5.7 m
Bedrock 0.4 m+

LOG

Geological classification	Lithology	Thickness m	Depth m
	Soil	0.5	0.5
Till	Clay, grey and brown, sandy, with pebbles of sandstone and limestone and a 0.6-m layer of fine sand	5.2	5.7
Upper Magnesian Limestone	Limestone, buff, dolomitic	0.4+	6.1

SE 35 NE 18 3758 5945 The Hollies, Knaresborough Outer **Block A**

Surface level (+52.1 m) +171 ft
Water struck at (+39.9 m) +131 ft
203 mm
March 1979

Waste 10.8 m
Bedrock 2.3 m+

LOG

Geological classification	Lithology	Thickness m	Depth m
	Soil	0.2	0.2
Till	Clay, brown, orange-brown and grey, silty and sandy with fine to coarse pebbles of dolomitic limestone, Carboniferous Limestone, sandstone, shale and quartzite	10.6	10.8
Upper Magnesian Limestone	Limestone, white, dolomitic, friable	2.3+	13.1

SE 35 NE 19 3851 5951 Clareton Moor Farm, Arkendale **Block A**

Surface level (+51.5 m) +169 ft
Water struck at (+46.1 m) +151 ft
203 mm
March 1979

Waste 17.3 m
Bedrock 0.5 m+

LOG

Geological classification	Lithology	Thickness m	Depth m
	Soil	0.3	0.3
Till	Clay, orange, brown and grey mottled, silty and sandy, pebbly, with rootlets; angular and subangular sandstone and quartzite cobbles and fine and coarse pebbles of sandstone and limestone	5.1	5.4
	Clay, very sandy, becoming silty and laminated from 8.2 to 10.2 m, and then pebbly with pebbles of Carboniferous Limestone, sandstone and dolomitic limestone, with some siltstone and quartzite	11.9	17.3
Upper Magnesian Limestone	Limestone, fresh, greyish white, dolomitic with manganese dendrites and small vugs	0.5+	17.8

SE 35 NE 20 **3975 5932** **Brunton Lane, Clareton** **Block A**

Surface level (+55.7 m) +183 ft
Water not encountered
203 mm
April 1979

Waste 10.7 m+

LOG

Geological classification	Lithology	Thickness m	Depth m
	Made ground and soil	0.7	0.7
Till	Clay, greyish brown, silty, with pebbles, cobbles and boulders of yellow-weathering arkosic sandstone and grey and black Carboniferous Limestone. The limestone cobbles and boulders are often fossiliferous and sometimes striated	10.0+	10.7
	Borehole abandoned because of drilling difficulties		

SE 35 NE 21 3683 5813 Hall Farm, Knaresborough Outer **Block A**

Surface level (+44.4 m) +146 ft
Water struck at (+37.9 m) +124 ft
203 mm
February 1979

Overburden 6.5 m
Mineral 5.4 m
Waste 2.5 m
Mineral 1.8 m
Waste 3.5 m+

LOG

Geological classification	Lithology	Thickness m	Depth m
	Soil	0.2	0.2
Till	Clay, yellow-brown and grey, silty and sandy, with small pebbles of dolomitic limestone and sandstone	6.3	6.5
Fluvio-glacial and Older River Sand and Gravel and associated clay	**a** Gravel Gravel: coarse and fine, angular to subrounded sandstone, quartzite and black limestone, with some white dolomitic limestone Sand: medium and coarse, with fine, subangular and subrounded clear quartz with sandstone and black limestone	3.1	9.6
	b 'Clayey' sand: medium and fine with some coarse, subangular and subrounded quartz and black limestone	2.3	11.9
	Silt, and laminated clay, very silty, sometimes sandy	2.5	14.4
	c 'Very clayey' sand: fine and medium, mainly subangular to rounded clear and yellow quartz with some white quartzite and rare black limestone	1.8	16.2
	Clay, grey, stiff, very silty, laminated with thin partings of silt	3.5+	19.7

GRADING

	Mean for deposit *percentages*			Depth below surface (m)	*percentages*						
	Fines	Sand	Gravel		Fines	Sand			Gravel		
					$-\frac{1}{16}$	$+\frac{1}{16}-\frac{1}{4}$	$+\frac{1}{4}-1$	+1–4	+4–16	+16–64	+64
a	4	35	61	6.5–7.6	3	4	3	4	26	61	0
				7.6–8.6	3	1	20	25	32	19	0
				8.6–9.6	5	9	25	17	14	23	7
				Mean	4	5	16	15	24	35	2
b	18	81	1	9.6–10.6	16	37	46	1	0	0	0
				10.6–11.9	18	30	37	13	2	0	0
				Mean	18	33	40	8	1	0	0
c	22	78	0	14.4–16.2	22	45	33	0	0	0	0
a+b +c	13	60	27	Mean	13	24	28	8	11	15	1

SE 35 NE 22 3723 5848 Hopewell House, Knaresborough Outer **Block A**

Surface level (+47.5 m) +156 ft
Water struck at (+39.0 m) +128 ft
203 mm
February 1979

Overburden 1.6 m
Mineral 6.3 m
Waste 0.6 m
Mineral 4.2 m
Waste 1.7 m
Mineral 2.7 m
Waste 0.2 m
Bedrock 1.7 m+

LOG

Geological classification	Lithology	Thickness m	Depth m
	Soil	0.4	0.4
Till	Clay, yellow and brown, sandy, with small pebbles of sandstone	1.2	1.6
Glacial Sand and Gravel	**a** 'Clayey' gravel Gravel: coarse and fine, with some cobbles, subangular and subrounded sandstone and quartzite with Carboniferous Limestone and some siltstone Sand: fine with medium and coarse subangular and subrounded clear and yellow-brown quartz, with some black limestone, white quartzite and white dolomitic limestone	6.3	7.9
Fluvio-glacial and Older River Sand and Gravel and associated clay	Clay, yellow-brown, streaked with grey, soft, laminated, very silty becoming sandy	0.6	8.5
	b 'Clayey' sand: medium and fine, with some coarse from 8.5 to 10.5 m, subrounded and subangular clear with yellow-brown quartz, with some white quartz and black limestone. Occasional fine pebbles of subrounded black limestone and white limestone down to 10.5 m	4.2	12.7
	Clay, brown, laminated, silty, micaceous	0.2	12.9
	Sand: medium and coarse, with fine, mainly subangular and subrounded quartz with black and white limestone and some white quartzite	0.6	13.5
	Clay, grey, firm, laminated with thin partings of brown micaceous silt	0.9	14.4
	c 'Clayey' gravel Gravel: coarse with fine, subangular and subrounded sandstone, and quartzite and siltstone pebbles Sand: medium and fine, with coarse, quartz with rare black limestone	2.7	17.1
	Clay, red-brown, with occasional green and grey mottles, silty and pebbly, with pebbles of red-brown marlstone and sandstone	0.2	17.3
Middle Marl	Marl, red-brown, with green mottles, micaceous, silty marl and clay	1.7+	19.0

GRADING

	Mean for deposit *percentages*			Depth below surface (m)	*percentages*						
	Fines	Sand	Gravel		Fines	Sand			Gravel		
					$-\frac{1}{16}$	$+\frac{1}{16}-\frac{1}{4}$	$+\frac{1}{4}-1$	$+1-4$	$+4-16$	$+16-64$	$+64$
a	17	26	57	1.6–2.6	18	13	6	7	17	28	11
				2.6–3.6	17	11	9	10	17	36	0
				3.6–4.6	19	8	7	12	24	29	0
				4.6–5.6	16	9	6	8	18	29	14
				5.6–7.9	16	11	6	9	20	33	5
				Mean	17	11	7	9	19	32	6
b	15	82	3	8.5–9.5	16	23	38	13	10	1	0
				9.5–10.5	15	32	45	7	1	0	0
				10.5–11.5	21	73	6	0	0	0	0
				11.5–12.7	9	76	14	0	0	0	0
				Mean	15	52	25	5	3	0	0
c	11	38	51	14.4–15.6	9	13	10	8	25	35	0
				15.6–16.6	15	18	20	7	10	25	5
				16.6–17.1	9	2	30	11	10	26	13
				Mean	11	12	17	8	17	30	4
a+b +c	15	47	38	Mean	15	24	15	8	13	21	4

SE 35 NE 23 3898 5889 Coneythorpe **Block A**

Surface level (+45.4 m) +149 ft
Water struck at (+42.6 m) +140 ft
203 mm
March 1979

Overburden 0.9 m
Mineral 4.0 m
Waste 16.8 m+

LOG

Geological classification	Lithology	Thickness m	Depth m
	Soil, dark brown, passing down into sandy clay and	0.9	0.9
Sand, Silt and Clay, Glacial Lake Deposits	'Clayey' sand, yellow Sand: fine with some medium, becoming finer with depth; mainly quartz with a trace of black limestone	4.0	4.9
	Silt, grey, very clayey, with some fine sand; becoming consolidated with depth and	4.1	9.0
	Clay, grey, firm, laminated, very silty, with partings of fine and medium yellow-brown quartz sand	0.9	9.9
Till	Clay, silty, pebbly, grey, becoming red-brown below 17.0 m; fine and coarse pebbles, and occasional cobbles, of quartzite, sandstone, limestone and some shale	8.0	17.9
Clay associated with Fluvio-glacial and Older River Sand and Gravel	Clay, brown and grey, very silty, finely laminated	3.8+	21.7

GRADING

Mean for deposit *percentages*			Depth below surface (m)	*percentages*						
Fines	Sand	Gravel		Fines	Sand			Gravel		
				$-\frac{1}{16}$	$+\frac{1}{16}-\frac{1}{4}$	$+\frac{1}{4}-1$	+1–4	+4–16	+16–64	+64
14	85	1	0.9–1.9	25	58	16	1	0	0	0
			1.9–2.9	14	79	7	0	0	0	0
			2.9–4.9	9	85	3	2	1	0	0
			Mean	14	77	7	1	1	0	0

SE 35 NE 24 3966 5855 Spring Bank Farm, Flaxby **Block A**

Surface level (+61.1 m) +200 ft
Water struck at (+47.9 m) +157 ft
203 mm
March 1979

Overburden 0.4 m
Mineral 22.4 m
Waste 1.0 m+

LOG

Geological classification	Lithology	Thickness m	Depth m
	Soil	0.4	0.4
Glacial Sand and Gravel	**a** 'Clayey' gravel Gravel: coarse and fine, angular to subrounded Carboniferous Limestone with sandstone and siliceous siltstone and some dolomitic limestone; trace of concretionary ironstone Sand: fine, medium and coarse quartz and black limestone	15.2	15.6
	b Gravel Gravel: coarse and fine, angular to subrounded Carboniferous Limestone and yellow and white arkosic sandstone Sand: coarse with medium and fine; quartz, black limestone and white sandstone	3.0	18.6
	c Pebbly sand Gravel: fine, angular to subrounded Carboniferous Limestone and Sandstone Sand: medium, coarse and fine, quartz, black limestone and sandstone	3.2	22.8
Clay associated with Fluvio-glacial and Older River Sand and Gravel	Clay, grey and brown, very silty; becoming a sandy silt towards 23.8 m	1.0+	23.8

GRADING

	Mean for deposit percentages: Fines	Sand	Gravel	Depth below surface (m)	percentages: Fines	Sand			Gravel		
					$-\frac{1}{16}$	$+\frac{1}{16}-\frac{1}{4}$	$+\frac{1}{4}-1$	+1–4	+4–16	+16–64	+64
a	18	38	44	0.4–1.4	11	16	10	12	19	30	3
				1.4–2.4	15	15	15	14	20	21	0
				2.4–3.4	23	19	13	10	15	19	2
				3.4–4.4	25	20	17	13	16	8	0
				4.4–5.4	34	36	14	9	7	0	0
				5.4–6.4	23	13	9	16	22	17	0
				6.4–7.4	15	14	11	14	22	25	0
				7.4–8.4	14	15	9	7	21	34	0
				8.4–9.4	18	11	9	12	21	27	2
				9.4–10.4	18	15	9	8	19	25	6
				10.4–11.4	20	19	10	6	16	30	0
				11.4–12.4	14	13	7	5	15	34	11
				12.4–13.2	16	17	9	9	18	29	3
				13.2–14.2	7	8	5	9	22	46	3
				14.2–15.6	20	17	12	11	14	27	0
				Mean	18	17	11	10	17	25	2
b	4	18	78	15.6–16.6	4	3	4	6	30	54	0
				16.6–17.6	5	1	5	22	32	32	3
				17.6–18.6	4	3	3	7	20	57	7
				Mean	4	2	4	12	27	48	3
c	5	75	20	18.6–19.6	3	8	32	50	6	0	0
				19.6–20.6	5	28	32	23	8	3	0
				20.6–21.6	6	23	36	24	10	1	0
				21.6–22.8	7	20	17	11	33	12	0
				Mean	5	20	29	26	15	5	0
a+b +c	14	42	44	Mean	14	15	13	14	18	24	2

COMPOSITION

	Depth below surface (m)	Percentage by weight in +4 mm fraction: Limestone	Dolomitic limestone	Sandstone	Quartz	Siltstone	Ironstone
a	8.4–9.4	69	7	23	0	0	1
	14.2–15.6	56	9	29	1	3	2

SE 35 NE 25 3689 5748 Hay-a-Park Lane, Knaresborough Outer **Block A**

Surface level (+43.5 m) +143 ft
Water struck at (+38.9 m) +128 ft
203 mm
February 1979

Overburden 4.8 m
Mineral 2.0 m
Waste 1.0 m
Mineral 1.0 m
Waste 1.0 m
Mineral 1.0 m
Waste 8.4 m
Mineral 4.4 m+

LOG

Geological classification	Lithology	Thickness m	Depth m
	Soil	0.6	0.6
Till	Clay, red and yellow-brown, sandy, pebbly; pebbles are subangular and subrounded dolomitic limestone, Carboniferous Limestone and sandstone, with occasional subangular sandstone cobbles	4.2	4.8
Fluvio-glacial and Older River Sand and Gravel and associated clay	**a** 'Clayey' pebbly sand: fine and medium, with some coarse; quartz, sandstone and black limestone	2.0	6.8
	Silt, sandy	1.0	7.8
	b 'Clayey' sand: fine and medium; quartz, with some sandstone and black limestone	1.0	8.8
	Silt, sandy	1.0	9.8
	c 'Very clayey' sand: fine, with some medium; quartz with some sandstone and black limestone	1.0	10.8
	Clay, silty, sandy, laminated with silt partings below 17.5 m	8.4	19.2
	d Gravel Gravel: coarse and fine, subangular to rounded Carboniferous Limestone and Magnesian Limestone and sandstone Sand: medium, coarse and fine; angular and subangular quartz and white and black limestone	4.4+	23.6

GRADING

	Mean for deposit *percentages*			Depth below surface (m)	*percentages*						
	Fines	Sand	Gravel		Fines	Sand			Gravel		
					$-\frac{1}{16}$	$+\frac{1}{16}-\frac{1}{4}$	$+\frac{1}{4}-1$	+1–4	+4–16	+16–64	+64
a	15	80	5	4.8–6.8	15	53	22	5	5	0	0
b	12	85	3	7.8–8.8	12	49	33	3	3	0	0
c	25	75	0	9.8–10.8	25	71	4	0	0	0	0
d	6	28	66	19.2–19.8	13	10	26	11	21	19	0
				19.8–20.8	7	7	6	6	30	39	5
				20.8–21.8	6	8	8	5	18	56	0
				21.8–22.6	3	4	18	7	15	54	0
				22.6–23.2	3	3	29	16	20	20	9
				23.2–23.6	1	5	7	6	13	66	2
				Mean	6	6	14	8	21	43	2
a+b+c+d	11	53	36	Mean	11	30	17	6	12	23	1

COMPOSITION

	Depth below surface (m)	*Percentage by weight in +4 mm fraction*					
		Limestone	Dolomitic limestone	Sandstone	Quartz	Siltstone	Ironstone
d	22.6–23.2	23	22	45	3	7	trace

SE 35 NE 26 3794 5770 Oakwood Farm, Knaresborough Outer **Block A**

Surface level (+41.8 m) +137 ft
Water struck at (+40.3 m) +132 ft
203 mm
February 1979

Waste 24.0 m
Bedrock 0.5 m+

LOG

Geological classification	Lithology	Thickness m	Depth m
Silt and Clay of the Glacial Lake Deposits	Soil on clay	0.6	0.6
Till	Clay, grey, sandy, pebbly, with pebbles and cobbles of black crinoidal limestone and grey and brown sandstone	7.2	8.0
Fluvio-glacial and Older River Sand and Gravel and associated clay	Clay and silt, grey and brown, laminated, with a sandy silt layer from 13.4 to 16.0 m	13.5	21.5
	Sandy gravel Gravel: coarse and fine, subangular to subrounded sandstone, dolomitic limestone and black limestone with some siliceous siltstone Sand: Coarse and medium, with fine; mainly quartz with some black limestone and white quartzite	2.5	24.0
Middle Marl	Clay, red-brown, very silty; weathered marl	0.5+	24.5

GRADING

	Mean for deposit *percentages*			Depth below surface (m)	*percentages*						
	Fines	Sand	Gravel		Fines	Sand			Gravel		
					$-\frac{1}{16}$	$+\frac{1}{16}-\frac{1}{4}$	$+\frac{1}{4}-1$	+1–4	+4–16	+16–64	+64
*	10	50	40	21.5–22.8	12	6	14	19	24	25	0
				22.8–23.5	9	17	53	12	4	5	0
				23.5–24.0	6	12	14	10	20	35	2
				Mean	10	10	25	15	18	21	1

* Non-mineral: excessive overburden.

COMPOSITION

Depth below surface (m)	*Percentage by weight in +4 mm fraction*					
	Limestone	Dolomitic limestone	Sandstone	Quartz	Siltstone	Ironstone
21.5–22.8	21	27	41	5	6	trace

SE 35 NE 27 3851 5718 York Road, Goldsborough **Block A**

Surface level (+45.1 m) +148 ft
Water struck at (+36.8 m) +121 ft
203 mm
February 1979

Waste 24.5 m+

LOG

Geological classification	Lithology	Thickness m	Depth m
	Soil	0.2	0.2
Till	Clay, reddish brown, becoming greyish brown below 3.0 m, silty and sandy, with pebbles of limestone and quartzite	6.8	7.0
Clay associated with Fluvio-glacial and Older River Sand and Gravel	Clay and silt, often laminated and sandy, with a pebbly sand layer from 9.1 to 10.6 m	17.5+	24.5

GRADING

	Mean for deposit *percentages*			Depth below surface (m)	*percentages*						
	Fines	Sand	Gravel		Fines	Sand			Gravel		
					$-\frac{1}{16}$	$+\frac{1}{16}-\frac{1}{4}$	$+\frac{1}{4}-1$	+1–4	+4–16	+16–64	+64
*	10	84	6	9.1–10.1	10	25	47	9	9	0	0
				10.1–10.6	9	19	67	5	0	0	0
				Mean	10	23	53	8	6	0	0

* Non-mineral: excessive overburden.

SE 35 NE 28 3959 5758 Flaxby **Block A**

Surface level (+41.9 m) +137 ft
Water struck at (+38.7 m) +127 ft
203 mm
March 1979

Overburden 4.1 m
Mineral 1.4 m
Waste 14.7 m+

LOG

Geological classification	Lithology	Thickness m	Depth m
	Soil	0.3	0.3
Till	Clay, yellow and brown, sandy, with occasional small angular quartzite pebbles	0.7	1.0
Fluvio-glacial and Older River Sand and Gravel and associated clay	Clay, yellowish brown, streaked with grey, silty, laminated	3.1	4.1
	'Very clayey' sand: fine with medium; mainly quartz, with some black limestone and siltstone	1.4	5.5
	Clay, brown and grey, laminated, silty, sometimes sandy	13.7	19.2
	Silt, yellowish brown, clayey and sandy	1.0+	20.2

GRADING

Mean for deposit *percentages*			Depth below surface (m)	*percentages*						
Fines	Sand	Gravel		Fines	Sand			Gravel		
				$-\frac{1}{16}$	$+\frac{1}{16}-\frac{1}{4}$	$+\frac{1}{4}-1$	+1–4	+4–16	+16–64	+64
23	77	0	4.1–5.5	23	58	17	2	0	0	0

SE 35 NE 29 3630 5604 Plompton Mill Farm, Knaresborough Block B

Surface level (+43.2 m) +142 ft
Water struck at (+32.2 m) +106 ft
152 mm
February 1979

Overburden 10.7 m
Mineral 3.7 m
Bedrock 3.0 m+

LOG

Geological classification	Lithology	Thickness m	Depth m
	Soil	0.2	0.2
Fluvio-glacial Terrace Deposits	Clay, brown and reddish brown, with numerous pebbles of brown sandstone and black crinoidal limestone; contains three layers (0.7, 0.8 and (**a**) 1.3 m thick) of 'very clayey' sand	10.5	10.7
	b 'Very clayey' pebbly sand Gravel: fine and coarse subangular to rounded brown sandstone and dolomitic limestone Sand: fine, medium and coarse, mainly quartz	3.7	14.4
Lower Magnesian Limestone	Dolomitic limestone	3.0+	17.4

GRADING

	Mean for deposit percentages			Depth below surface (m)	percentages						
	Fines	Sand	Gravel		Fines	Sand			Gravel		
					$-\frac{1}{16}$	$+\frac{1}{16}-\frac{1}{4}$	$+\frac{1}{4}-1$	+1–4	+4–16	+16–64	+64
a*	35	65	0	9.0–10.3	35	37	27	1	0	0	0
b	25	57	18	10.7–11.7	23	22	18	13	13	0	0
				11.7–12.7	22	15	27	23	7	6	0
				12.7–14.4	28	12	23	20	11	6	0
				Mean	25	16	22	19	10	8	0
a+b	28	59	13	Mean	28	22	23	14	7	6	0

* Non-mineral: excessive overburden.

SE 35 NE 30 **3675 5661** **Haughs Farm, Knaresborough** **Block B**

Surface level (+32.5 m) +107 ft
Water struck at (+28.5 m) +94 ft
203 mm
January 1979

Overburden 4.0 m
Mineral 2.1 m
Waste 7.6 m
Bedrock 1.7 m+

LOG

Geological classification	Lithology	Thickness m	Depth m
Made ground	Fill	0.3	0.3
Alluvium	Clay, yellow-brown passing down into grey; very silty and sandy	3.7	4.0
	Sandy gravel Gravel: coarse and fine, rounded and subrounded black limestone and grey sandstone and quartzite Sand: medium, with coarse and fine, angular and subangular quartz and black limestone	2.1	6.1
Till?	Clay, red-brown and grey, marly, with subangular pebbles of red-brown siltstone, yellow siltstone and yellow and white sandstone; occasional cobbles of fibrous gypsum and limestone	7.6	13.7
Middle Marl	Marl and siltstone, grey-green, with some red brown, weak, highly weathered, micaceous	1.7+	15.4

GRADING

Mean for deposit *percentages*			Depth below surface (m)	*percentages*						
Fines	Sand	Gravel		Fines	Sand			Gravel		
				$-\frac{1}{16}$	$+\frac{1}{16}-\frac{1}{4}$	$+\frac{1}{4}-1$	$+1-4$	$+4-16$	$+16-64$	$+64$
5	61	34	4.0–5.0	7	15	27	6	14	32	0
			5.0–6.1	3	4	59	11	5	18	0
			Mean	5	9	43	9	9	25	0

SE 35 NE 31 3798 5670 Cockstone Hill, Goldsborough **Block A**

Surface level (+45.7 m) +150 ft
Water struck at (+36.8 m) +121 ft
203 mm
January 1979

Overburden 8.9 m
Mineral 3.7 m
Waste 2.6 m
Bedrock 0.8 m

LOG

Geological classification	Lithology	Thickness m	Depth m
	Soil	0.4	0.4
Till	Clay, yellow-brown, passing down into grey and brown, sandy and pebbly with occasional cobbles. Pebbles mainly sandstone	8.0	8.4
Fluvio-glacial and Older River Sand and Gravel and associated clay	Clay, brown, firm, laminated, with partings of fine silt	0.5	8.9
	'Clayey' sand, with some pebbles of yellow and white limestone Sand: medium and fine with some coarse, angular and subangular quartz, with some white and black limestone	3.7	12.6
	Clay, yellow-brown, very silty, finely laminated with partings of brown micaceous silt	2.6	15.2
Upper Magnesian Limestone	Limestone, dolomitic, greyish-white, fresh, with small vugs	0.8+	16.0

GRADING

Mean for deposit *percentages*			Depth below surface (m)	*percentages*						
Fines	Sand	Gravel		Fines	Sand			Gravel		
				$-\frac{1}{16}$	$+\frac{1}{16}-\frac{1}{4}$	$+\frac{1}{4}-1$	+1–4	+4–16	+16–64	+64
14	83	3	8.9–9.9	17	34	41	4	4	0	0
			9.9–11.7	7	11	73	6	3	0	0
			11.7–12.6	26	70	4	0	0	0	0
			Mean	14	31	48	4	3	0	0

SE 35 NE 32 3972 5656 Goldsborough Moor **Block A**

Surface level (+30.5 m) +100 ft
Water struck at (+24.0 m) +79 ft, artesian
203 mm
April 1979

Overburden 0.5 m
Mineral 1.0 m
Waste 13.7 m
Mineral 4.6 m
Bedrock 1.4 m+

LOG

Geological classification	Lithology	Thickness m	Depth m
	Soil, dusky brown silty clay	0.2	0.2
Silt and Clay of the Glacial Lake Deposits	Clay, greyish-brown silty	0.3	0.5
Glacial Sand and Gravel	**a** 'Clayey' sandy gravel Gravel: fine and coarse, subrounded to rounded quartzitic sandstone, with black subangular limestone Sand: medium, fine and coarse; subangular quartz, quartzitic sandstone and black limestone	1.0	1.5
Till	Clay, sandy, with pebbles	0.2	1.7
Fluvio-glacial and Older River Sand and Gravel and associated clay	Clay, brown and greyish brown, silty, laminated, with a layer of 'very clayey' sand (**b**) from 8.2 to 10.0 m, and of 'very clayey' pebbly sand (**c**) from 12.8 to 14.5 m	13.5	15.2
	d Sandy gravel Gravel: fine and coarse, rounded and well rounded sandstone, siltstone and dolomitic limestone, with some black limestone Sand: medium, coarse and fine, mainly subangular quartz with subrounded limestone and subangular dolomitic limestone	4.6	19.8
Upper Marl?	Sand, red-brown, clayey, fine-grained	0.9	20.7
Upper Marl	Marl, dark reddish brown, with yellowish green mottles and laminae	0.5+	21.2

GRADING

	Mean for deposit *percentages*			Depth below surface (m)	*percentages*						
					Fines	Sand			Gravel		
	Fines	Sand	Gravel		$-\frac{1}{16}$	$+\frac{1}{16}-\frac{1}{4}$	$+\frac{1}{4}-1$	+1–4	+4–16	+16–64	+64
a	20	46	34	0.5–1.5	20	17	19	10	24	10	0
b*	33	67	0	8.2–9.0	32	56	12	0	0	0	0
				9.0–10.0	35	49	15	1	0	0	0
				Mean	33	52	14	1	0	0	0
c*	27	62	11	12.8–13.5	21	18	27	12	10	12	0
				13.5–14.5	31	49	15	2	3	0	0
				Mean	27	36	20	6	6	5	0
d	6	64	30	15.2–15.8	13	21	63	3	0	0	0
				15.8–16.8	7	16	38	5	16	16	3
				16.8–17.8	4	10	34	13	21	18	0
				17.8–18.8	4	6	36	15	21	19	0
				18.8–19.8	6	10	48	13	13	10	0
				Mean	6	12	42	10	16	13	1
a+b+ c+d	17	62	21	Mean	17	25	30	7	12	9	0

* Non-mineral: excessive overburden.

COMPOSITION

	Depth below surface (m)	*Percentage by weight in +4 mm fraction*					
		Limestone	Dolomitic limestone	Sandstone	Quartz	Siltstone	Ironstone
d	16.8–17.8	3	11	67	4	14	1

SE 35 NE 33 3753 5519 Tickhill Farm, Plompton **Block B**

Surface level (+30.5 m) +100 ft
Water not encountered
152 mm
March 1979

Waste 2.9 m
Bedrock 4.1 m+

LOG

Geological classification	Lithology	Thickness m	Depth m
	Soil	0.5	0.5
River Terrace Deposits, undifferentiated	Clay, yellow, very sandy, with pebbles and cobbles of sandstone	2.4	2.9
Middle Marl	Clay, red and brown, mottled green and grey, silty and marly	3.0	5.9
Lower Magnesian Limestone	Limestone, yellow-brown, dolomitic	1.1+	7.0

SE 35 NE 34 3760 5582 Avenue House Farm, Goldsborough **Block B**

Surface level (+39.7 m) +130 ft
Water struck at (+31.0 m) +102 ft
203 mm
January 1979

Overburden 0.4 m
Mineral 1.2 m
Waste 1.7 m
Mineral 1.9 m
Waste 2.2 m
Mineral 1.1 m
Waste 3.5 m+

LOG

Geological classification	Lithology	Thickness m	Depth m
	Soil	0.4	0.4
Fluvio-glacial Terrace Deposits	**a** 'Clayey' gravel Gravel: Coarse and fine, subangular to rounded dolomitic limestone, sandstone and quartzite with some limestone Sand: Fine to coarse, angular and subangular quartz, quartzitic sandstone and dolomitic limestone	1.2	1.6
Undifferentiated Silt and Clay of the Glacial Lake Deposits and Fluvio-glacial and Older River Sand and Gravel and associated clay	Clay, grey and brown, silty, laminated, with partings of silt and fine sand	1.7	3.3
	b 'Clayey' sand: fine and medium with some coarse; subangular and subrounded quartz with some black limestone	1.9	5.2
	Clay, grey, stiff, laminated with partings of greyish brown micaceous silt	2.2	7.4
	c 'Very clayey' sand: fine and medium; mainly subangular and subrounded quartz	1.1	8.5
	Clay, grey, silty, laminated, with partings of fine and medium sand passing down into 'very clayey' sand	0.7	9.2
Sandy Till?	Sandy pebbly clay, red-brown, containing pebbles of coarse sandstone, Carboniferous Limestone, and, below 10.4 m, gypsum	2.8+	12.0

GRADING

	Mean for deposit *percentages*			Depth below surface (m)	*percentages*						
	Fines	Sand	Gravel		Fines	Sand			Gravel		
					$-\frac{1}{16}$	$+\frac{1}{16}-\frac{1}{4}$	$+\frac{1}{4}-1$	+1–4	+4–16	+16–64	+64
a	13	28	59	0.4–1.6	13	11	9	8	15	35	9
b	11	88	1	3.3–4.3	6	37	49	8	0	0	0
				4.3–5.2	17	26	52	3	2	0	0
				Mean	11	32	51	5	1	0	0
c	39	61	0	7.4–8.5	39	48	12	1	0	0	0
a+b +c	19	64	17	Mean	21	32	27	4	5	9	2

SE 35 NE 35 3863 5576 Goldsborough Park **Block A**

Surface level (+30.3 m) +99 ft
Water struck at (+25.0 m) +82 ft, artesian
203 mm
April 1979

Waste 13.9 m
Bedrock 0.1 m+

LOG

Geological classification	Lithology	Thickness m	Depth m
	Soil	0.2	0.2
Undifferentiated Silt and Clay of the Glacial Lake Deposits and clay associated with Fluvio-glacial and Older River Sand and Gravel	Clay, dark yellowish brown, laminated, with silty partings	5.1	5.3
	Sandy silt	1.5	6.8
	Clay, greyish brown, laminated, silty	0.6	7.4
	Sandy silt	1.0	8.4
	Clay, greyish brown, laminated, silty, with numerous weathered pebbles of Magnesian Limestone near the base	2.2	10.6
	'Very clayey' sand, reddish brown, and sandy silt becoming semi-consolidated and micaceous near the base	3.3	13.9
Upper Marl	Marl, red, clayey	0.1+	14.0

GRADING

	Mean for deposit *percentages*			Depth below surface (m)	*percentages*						
	Fines	Sand	Gravel		Fines	Sand			Gravel		
					$-\frac{1}{16}$	$+\frac{1}{16}-\frac{1}{4}$	$+\frac{1}{4}-1$	+1–4	+4–16	+16–64	+64
*	38	62	0	11.3–12.3	38	58	3	1	0	0	0

* Non-mineral: excessive overburden.

SE 35 NE 36 3961 5510 Ribston Lodge, Great Ribston **Block A**

Surface level (+37.4 m) +123 ft
Water struck at (+32.4 m) +106 ft
152 mm
April 1979

Overburden 0.6 m
Mineral 4.4 m
Waste 2.0 m
Mineral 2.0 m
Waste 13.0 m+

LOG

Geological classification	Lithology	Thickness m	Depth m
	Soil	0.6	0.6
Glacial Sand and Gravel	**a** 'Very clayey' pebbly sand Gravel: fine and coarse pebbles of yellow dolomitic limestone, and grey and black limestone Sand: fine and medium, with some coarse; mainly angular to subrounded quartz with some limestone	4.4	5.0
Till	Silt, pale brown, micaceous	2.0	7.0
Glacial Sand and Gravel	**b** 'Very clayey' sand: fine and medium, mainly subangular to subrounded quartz with some limestone	2.0	9.0
Clay associated with Fluvio-glacial and Older River Sand and Gravel	Silt and laminated clay	13.0+	22.0

GRADING

	Mean for deposit *percentages*			Depth below surface (m)	*percentages*						
	Fines	Sand	Gravel		Fines	Sand			Gravel		
					$-\frac{1}{16}$	$+\frac{1}{16}-\frac{1}{4}$	$+\frac{1}{4}-1$	$+1-4$	$+4-16$	$+16-64$	$+64$
a	28	68	4	0.6–1.0	21	18	11	11	21	17	0
				1.0–2.0	36	42	13	6	2	1	0
				2.0–3.0	19	63	16	2	0	0	0
				3.0–4.0	33	52	13	2	0	0	0
				4.0–5.0	26	52	19	3	0	0	0
				Mean	28	49	15	4	2	2	0
b	34	66	0	7.0–8.0	36	48	16	0	0	0	0
				8.0–9.0	32	33	35	0	0	0	0
				Mean	34	41	25	0	0	0	0
a+b	30	67	3	Mean	30	46	18	3	2	1	0

SE 35 SE 16 3739 5456 Scaliber Farm, Plompton **Block B**

Surface level (+34.9 m) +115 ft
Water struck at (+28.9 m) +95 ft
152 mm
February 1979

Overburden 4.6 m
Mineral 2.5 m
Waste 1.4 m
Bedrock 1.7 m+

LOG

Geological classification	Lithology	Thickness m	Depth m
	Soil	0.3	0.3
Till	Clay, brown, sandy with pebbles of yellow sandstone, and with fine sand partings	4.3	4.6
Glacial Sand and Gravel	'Very clayey' pebbly sand Gravel: fine and coarse angular to subrounded sandstone and limestone Sand: fine to coarse subrounded quartz, with sandstone and limestone	2.5	7.1
	Clay, yellow, ochreous with fragments of brown marl	1.4	8.5
Lower Magnesian Limestone	Limestone, yellow, weathered, dolomitic	1.7+	10.2

GRADING

Mean for deposit *percentages*			Depth below surface (m)	*percentages*						
Fines	Sand	Gravel		Fines	Sand			Gravel		
				$-\frac{1}{16}$	$+\frac{1}{16}-\frac{1}{4}$	$+\frac{1}{4}-1$	+1–4	+4–16	+16–64	+64
27	58	15	4.6–5.1	34	32	28	6	0	0	0
			5.1–6.1	33	16	30	21	0	0	0
			6.1–7.1	19	7	19	19	21	16	0
			Mean	27	16	25	17	9	6	0

SE 35 SE 17 **3822 5441** **Low Field, Goldsborough** **Block B**

Surface level (+26.7 m) +88 ft
Water struck at (+24.0 m) +79 ft
203 mm
April 1979

Overburden 6.4 m
Mineral 2.2 m
Bedrock 1.3 m+

LOG

Geological classification	Lithology	Thickness m	Depth m
	Soil	0.2	0.2
River Terrace Deposits, undifferentiated	Clayey silt and sandy clay, yellowish brown	2.6	2.8
	a 'Clayey' gravel Gravel: coarse and fine, subrounded brown sandstone and black limestone with some white quartzite and a trace of weathered dolomitic limestone Sand: fine to coarse, quartz, quartzite and grey limestone	0.7	3.5
Fluvio-glacial and Older River Sand and Gravel and associated clay	Clay, greyish-brown, silty, laminated	0.5	4.0
	b 'Very clayey' sand: fine with some medium, subrounded quartz and subangular Magnesian Limestone	0.8	4.8
	Clay, greyish brown, silty, laminated	0.2	5.0
	c 'Clayey' sand, fine with some medium, quartz with some dolomitic limestone	1.0	6.0
	Clay, greyish brown, silty, laminated	0.4	6.4
	d 'Clayey' gravel Gravel: fine and coarse mainly subangular yellow dolomitic limestone and white quartzite, with some quartzite sandstone Sand: coarse, medium and fine, quartz and dolomitic limestone	2.2	8.6
Middle Marl	Marl, red, silty	1.3+	9.9

GRADING

	Mean for deposit *percentages*			Depth below surface (m)	*percentages*						
	Fines	Sand	Gravel		Fines	Sand			Gravel		
					$-\frac{1}{16}$	$+\frac{1}{16}-\frac{1}{4}$	$+\frac{1}{4}-1$	$+1-4$	$+4-16$	$+16-64$	$+64$
a*	15	34	51	2.8–3.5	15	6	17	11	24	27	0
b*	39	60	1	4.0–4.8	39	50	9	1	1	0	0
c†	19	81	0	5.0–6.0	19	70	10	1	0	0	0
d	14	39	47	6.4–7.3	16	8	15	17	26	16	2
				7.3–8.6	13	7	15	16	28	22	0
				Mean	14	7	15	17	27	19	1
a+b+c+d	19	51	30	Mean	19	28	13	10	17	13	0

* Non-mineral: less than 1 m thick.
† Non-mineral: excessive overburden.

SE 35 SE 18 3864 5454 Gundrifs Beck, Goldsborough **Block B**

Surface level (+27.3 m) +90 ft
Water struck at (+25.5 m) +84 ft
203 mm
April 1979

Waste 8.4 m
Bedrock 1.4 m+

LOG

Geological classification	Lithology	Thickness m	Depth m
	Made ground and soil	0.4	0.4
River Terrace Deposits, undifferentiated	Clay, yellowish-brown, sandy, with quartzitic sandstone pebbles	1.4	1.8
	a Gravel Gravel: coarse and fine, with pebbles of subrounded and rounded quartzitic sandstone, black subangular to subrounded limestone, with some yellowish dolomitic limestone and white quartzite Sand: medium and coarse, with some fine, mainly quartz, sandstone and limestone	0.9	2.7
Fluvio-glacial and Older River Sand and Gravel and associated clay	Clay, greyish brown, silty, laminated	3.7	6.4
	b 'Clayey' sandy gravel, with two thin clay bands Gravel: fine and coarse, subangular pebbles of yellowish dolomitic limestone, and subrounded reddish quartzitic sandstone, with a trace of white quartzite Sand: coarse and medium, with some fine; mainly subangular dolomitic limestone and quartz, with quartzite	2.0	8.4
Sherwood Sandstone Group	Sand, reddish brown, becoming consolidated and passing down into friable sandstone	1.4+	9.8

GRADING

	Mean for deposit *percentages*			Depth below surface (m)	*percentages*						
	Fines	Sand	Gravel		Fines	Sand			Gravel		
					$-\frac{1}{16}$	$+\frac{1}{16}-\frac{1}{4}$	$+\frac{1}{4}-1$	+1–4	+4–16	+16–64	+64
a*	8	34	48	1.8–2.7	8	4	22	18	29	19	0
b†	14	51	35	6.4–8.4	14	8	28	15	20	8	7

* Non-mineral: less than 1 m thick.
† Non-mineral: excessive overburden.

SE 35 SE 19 3938 5448 Ribston Park, Great Ribston **Block A**

Surface level (+34.2 m) +112 ft
Water struck at (+31.7 m) +104 ft
152 mm
March 1979

Overburden 3.0 m
Mineral 3.0 m
Waste 1.0 m
Mineral 1.0 m
Waste 1.0 m
Mineral 1.5 m+

LOG

Geological classification	Lithology	Thickness m	Depth m
	Soil, sandy with pebbles	0.5	0.5
Till	Clay, dark brown, sandy with rounded pebbles of brown and grey sandstone and quartzite, passing down into 'very clayey' gravel and pebbly clay	2.5	3.0
Glacial Sand and Gravel with intercalated Till	**a** 'Very clayey' sand: fine, with medium; subangular to subrounded quartz, with quartzite and a trace of black limestone	3.0	6.0
	Sandy clay, pale brown	1.0	7.0
	b 'Very clayey' sand: fine with medium, pale buff; quartz, limestone and dolomitic limestone	1.0	8.0
	Sandy clay, pale brown	1.0	9.0
	c 'Clayey' sand: fine and medium, pale buff; quartz, limestone and dolomitic limestone	1.5+	10.5

GRADING

	Mean for deposit *percentages*			Depth below surface (m)	*percentages*						
	Fines	Sand	Gravel		Fines	Sand			Gravel		
					$-\frac{1}{16}$	$+\frac{1}{16}-\frac{1}{4}$	$+\frac{1}{4}-1$	+1–4	+4–16	+16–64	+64
a	24	76	0	3.0–5.0	23	70	7	0	0	0	0
				5.0–6.0	27	61	12	0	0	0	0
				Mean	24	67	9	0	0	0	0
b	33	67	0	7.0–8.0	33	50	16	1	0	0	0
c	13	87	0	9.0–10.5	13	54	32	1	0	0	0
a+b +c	23	77	0	Mean	23	60	17	0	0	0	0

SE 35 SE 20 3872 5310 Grange Farm, Little Ribston Block B

Surface level (+30.3 m) +99 ft
Water struck at (+28.3 m) +93 ft
152 mm
March 1979

Waste 15.5 m+

LOG

Geological classification	Lithology	Thickness m	Depth m
	Soil	0.6	0.6
Fluvio-glacial Terrace Deposits	Clay, brown silty	0.9	1.5
	a 'Clayey' gravel Gravel: coarse and fine, mainly angular to subrounded sandstone Sand: medium and coarse with some fine; mainly quartz	0.5	2.0
Clay associated with Fluvio-glacial and Older River Sand and Gravel	Clay, grey-brown, silty, laminated, with silt partings	10.6	12.6
Fluvio-glacial and Older River Sand and Gravel	**b** Gravel Gravel: coarse with fine; mainly subangular to rounded grey-brown fine-grained sandstone Sand: coarse and medium with some fine; subangular to subrounded quartz with trace of black limestone	2.9+	15.5

GRADING

	Mean for deposit *percentages*			Depth below surface (m)	*percentages*						
	Fines	Sand	Gravel		Fines	Sand			Gravel		
					$-\frac{1}{16}$	$+\frac{1}{16}-\frac{1}{4}$	$+\frac{1}{4}-1$	+1–4	+4–16	+16–64	+64
a*	13	38	49	1.5–2.0	13	5	20	13	20	29	0
b†	5	43	52	12.6–13.6	5	3	5	4	12	69	2
				13.6–14.6	5	6	33	29	9	16	2
				14.6–15.0	7	5	18	21	20	29	0
				15.0–15.5	5	4	30	22	11	28	0
				Mean	5	4	21	18	12	38	2
a+b	6	42	52	Mean	6	4	21	17	13	37	2

* Non-mineral: less than 1 m thick.
† Non-mineral: excessive overburden.

SE 35 SE 21 **3960 5366** **Ribston Hall, Great Ribston** **Block B**

Surface level (+26.0 m) +85 ft
Water struck at (+23.5 m) +77 ft
152 mm
April 1979

Overburden 0.7 m
Mineral 2.8 m
Waste 7.2 m
Bedrock 1.2 m+

LOG

Geological classification	Lithology	Thickness m	Depth m
	Soil, passing down into silty sand	0.7	0.7
Alluvium	'Clayey' gravel Gravel: coarse with fine; angular to subrounded siliceous sandstone and grey limestone Sand: fine and medium with some coarse, angular to rounded quartz with limestone	2.8	3.5
Till	Clay, grey, stiff, pebbly, with pebbles of brown and grey sandstone and black and grey limestone, with silty and gravelly layers	7.2	10.7
Sherwood Sandstone Group	'Clayey' sand, red, becoming consolidated into red friable flaggy sandstone below 11.7 m	1.2+	11.9

GRADING

Mean for deposit *percentages*			Depth below surface (m)	*percentages*						
Fines	Sand	Gravel		Fines	Sand			Gravel		
				$-\frac{1}{16}$	$+\frac{1}{16}-\frac{1}{4}$	$+\frac{1}{4}-1$	+1–4	+4–16	+16–64	+64
16	39	45	0.7–1.4	32	43	23	1	0	0	0
			1.4–3.5	11	9	14	5	15	46	0
			Mean	16	18	17	4	11	34	0

SE 35 SE 22 3990 5310 South Park, Little Ribston **Block B**

Surface level (+25.0 m) +82 ft
Water struck at (+22.8 m) +75 ft
152 mm
April 1979

Overburden 1.0 m
Mineral 1.2 m
Waste 5.7 m
Mineral 4.2 m+

LOG

Geological classification	Lithology	Thickness m	Depth m
	Soil	0.3	0.3
River Terrace Deposits, undifferentiated	Clay, dark brown silty	0.7	1.0
	a 'Clayey' sandy gravel Gravel: coarse and fine, rounded brown sandstone and white quartzite Sand: medium, with coarse and fine; mainly subangular to rounded quartz and sandstone	1.2	2.2
	Clay, grey, with occasional sandstone pebbles	0.2	2.4
	b 'Clayey' sandy gravel Gravel: coarse and fine, angular to rounded brown tabular sandstone Sand: fine to coarse, angular to subrounded quartz	0.6	3.0
Till	Clay, grey to black, with angular to rounded pebbles of sandstone and black limestone	4.9	7.9
Fluvio-glacial and Older River Sand and Gravel	**c** Sandy gravel Gravel: fine and coarse, red sandstone, quartzite, black limestone, and yellow dolomitic limestone and white quartz Sand: coarse and medium, with some fine; angular to subrounded quartz with black and yellow limestone	4.2+	12.1

GRADING

	Mean for deposit *percentages*			Depth below surface (m)	*percentages*						
	Fines	Sand	Gravel		Fines	Sand			Gravel		
					$-\frac{1}{16}$	$+\frac{1}{16}-\frac{1}{4}$	$+\frac{1}{4}-1$	+1–4	+4–16	+16–64	+64
a	19	56	25	1.0–2.2	19	13	33	10	8	17	0
b*	20	41	39	2.4–3.0	20	16	16	9	15	24	0
c	4	58	38	7.8–8.9	4	4	40	39	12	1	0
				8.9–9.9	4	5	21	30	29	11	0
				9.9–10.9	4	4	18	29	24	21	0
				10.9–12.1	2	0	12	36	22	25	3
				Mean	4	3	22	33	22	15	1
a+b +c	8	56	36	Mean	8	6	24	26	18	17	1

* Non-mineral: less than 1 m thick.

SE 35 SE 23 3870 5227 The Warrens, North Deighton **Block B**

Surface level (+30.2 m) +99 ft
Water struck at (+28.2 m) +93 ft
152 mm
March 1979

Overburden 0.1 m
Mineral 2.9 m
Waste 8.7 m
Bedrock 0.3 m+

LOG

Geological classification	Lithology	Thickness m	Depth m
	Soil	0.1	0.1
Fluvio-glacial Terrace Deposits	'Very clayey' sandy gravel Gravel: coarse and fine, rounded to well rounded brown sandstone, quartzite and black limestone Sand: medium, with fine and coarse; subangular to rounded quartz, quartzite and black limestone and brown	2.9	3.0
Clay associated with Fluvio-glacial and Older River Sand and Gravel	Clay, grey, silty, laminated, with pebbles and cobbles of yellow dolomitic limestone near the base	8.7	11.7
Lower Magnesian Limestone	Limestone, yellow, dolomitic	0.3+	12.0

GRADING

Mean for deposit *percentages*			Depth below surface (m)	*percentages*						
Fines	Sand	Gravel		Fines	Sand			Gravel		
				$-\frac{1}{16}$	$+\frac{1}{16}-\frac{1}{4}$	$+\frac{1}{4}-1$	+1–4	+4–16	+16–64	+64
25	45	30	0.1–1.3	40	18	27	7	6	2	0
			1.3–2.3	19	14	32	8	13	14	0
			2.3–3.0	6	2	13	10	23	45	0
			Mean	25	12	25	8	13	17	0

SE 35 SE 24 3960 5261 Sewage Works, Little Ribston **Block C**

Surface level (+30.5 m) +100 ft
Water struck at (+28.7 m) +94 ft
152 mm
March 1979

Waste 7.4 m
Bedrock 1.8 m+

LOG

Geological classification	Lithology	Thickness m	Depth m
	Soil	0.2	0.2
Till	Clay, dark brown, sandy, with pebbles and cobbles of sandstone and grey and black limestone	7.2	7.4
Upper Magnesian Limestone	Limestone, weathered, yellow, dolomitic	1.8+	9.2

SE 35 SE 25 3829 5181 Deighton Spring, North Deighton **Block B**

Surface level (+30.0 m) +98 ft
Water struck at (+28.0 m) +92 ft
152 mm
March 1979

Overburden 7.5 m
Mineral 4.0 m
Bedrock 0.5 m+

LOG

Geological classification	Lithology	Thickness m	Depth m
	Soil, slightly sandy clay	0.2	0.2
Silt and Clay of the Glacial Lake Deposits	Clay, brown, with pebbles of quartzite, sandstone and black limestone	5.1	5.3
Fluvio-glacial and Older River Sand and Gravel and associated clay	Clay, brown, silty, laminated	2.2	7.5
	Gravel	4.0	11.5
	Gravel: fine and coarse, mainly rounded sandstone and quartzite with some black limestone		
	Sand: coarse and medium; angular and subangular quartz, quartzite and black limestone		
Lower Magnesian Limestone	Limestone, yellow, dolomitic	0.5+	12.0

GRADING

Mean for deposit *percentages* Fines	Sand	Gravel	Depth below surface (m)	*percentages* Fines $-\frac{1}{16}$	Sand $+\frac{1}{16}-\frac{1}{4}$	$+\frac{1}{4}-1$	$+1-4$	Gravel $+4-16$	$+16-64$	$+64$
8	41	51	7.5–8.5	4	3	21	28	27	17	0
			8.5–9.5	16	1	7	13	41	22	0
			9.5–10.5	3	1	17	38	26	16	0
			10.5–11.5	10	2	10	24	35	19	0
			Mean	8	1	14	26	32	19	0

SE 35 SE 26 3960 5168 Dovecote Farm, North Deighton **Block C**

Surface level (+31.2 m) +102 ft
Water struck at (+29.2 m) +96 ft
152 mm
March 1979

Overburden 0.3 m
Mineral 1.0 m
Waste 9.2 m
Bedrock 0.4 m+

LOG

Geological classification	Lithology	Thickness m	Depth m
	Soil	0.3	0.3
Glacial Sand and Gravel	'Very clayey' sand: fine and medium; subangular to subrounded quartz with some black limestone	1.0	1.3
Till	Sandy clay, pale brown, pebbly, with pebbles of sandstone and black limestone	9.2	10.5
Lower Magnesian Limestone	Limestone, yellow and grey, dolomitic	0.4+	10.9

GRADING

Mean for deposit *percentages* Fines	Sand	Gravel	Depth below surface (m)	*percentages* Fines $-\frac{1}{16}$	Sand $+\frac{1}{16}-\frac{1}{4}$	$+\frac{1}{4}-1$	$+1-4$	Gravel $+4-16$	$+16-64$	$+64$
38	62	0	0.3–1.3	38	48	12	2	0	0	0

SE 35 SE 27 3983 5110 Woodlands Farm, Kirk Deighton **Block C**

Surface level (+34.6 m) +114 ft
Water not encountered
152 mm
March 1979

Waste 8.1 m
Bedrock 0.4 m+

LOG

Geological classification	Lithology	Thickness m	Depth m
	Soil	0.5	0.5
Till	Clay, grey-brown and greenish brown, with angular and subangular pebbles of sandstone, grey limestone, white dolomitic limestone and siltstone	5.0	5.5
Glacial Sand and Gravel	'Very clayey' gravel Gravel: fine to cobble-size; subangular to rounded grey limestone yellow dolomitic limestone, sandstone and quartzite Sand: medium, coarse and fine; subangular to subrounded quartz and limestone	0.6	6.1
Till	Clay, brown, sandy, with pebbles and cobbles of sandstone, limestone and dolomitic limestone, becoming reddish in the lowermost 0.6 m	1.4	8.1
Lower Magnesian Limestone	Limestone, yellow, weathered, dolomitic	0.4+	8.5

GRADING

	Mean for deposit *percentages* Fines	Sand	Gravel	Depth below surface (m)	*percentages* Fines $-\frac{1}{16}$	Sand $+\frac{1}{16}-\frac{1}{4}$	$+\frac{1}{4}-1$	$+1-4$	Gravel $+4-16$	$+16-64$	$+64$
*	22	38	40	5.5–6.1	22	9	18	11	17	23	0

* Non-mineral: less than 1 m thick, and overburden excessive.

SE 35 SE 28 **3633 5156** **Spofforth Mill** **Block B**

Surface level (+32.9 m) +108 ft
Water struck at (+29.9 m) +98 ft
152 mm
April 1979

Overburden 3.0 m
Mineral 1.5 m
Waste 2.6 m
Bedrock 1.0 m+

LOG

Geological classification	Lithology	Thickness m	Depth m
	Soil	0.2	0.2
Alluvium	Clay, brown and grey, sandy, with some sandstone pebbles	2.8	3.0
	Gravel Gravel: coarse and fine; subrounded to rounded brown sandstone, with some grey limestone and white quartzite Sand: medium and coarse, with some fine; angular to subangular quartz with some quartzite	1.5	4.5
Till	Clay, brown, with numerous pebbles of sandstone and grey limestone	2.6	7.1
Namurian, undivided	Mudstone, purple, with partings of finely disseminated mica	1.0+	8.1

GRADING

Mean for deposit *percentages*			Depth below surface (m)	*percentages*						
Fines	Sand	Gravel		Fines	Sand			Gravel		
				$-\frac{1}{16}$	$+\frac{1}{16}-\frac{1}{4}$	$+\frac{1}{4}-1$	+1–4	+4–16	+16–64	+64
6	44	50	3.0–4.5	6	4	20	20	18	32	0

APPENDIX G

CONVERSION TABLE, METRES TO FEET (to nearest 0.5 ft)

m	ft	m	ft	m	ft	m	ft	m	ft
0.1	0.5	6.1	20	12.1	39.5	18.1	59.5	24.1	79
0.2	0.5	6.2	20.5	12.2	40	18.2	59.5	24.2	79.5
0.3	1	6.3	20.5	12.3	40.5	18.3	60	24.3	79.5
0.4	1.5	6.4	21	12.4	40.5	18.4	60.5	24.4	80
0.5	1.5	6.5	21.5	12.5	41	18.5	60.5	24.5	80.5
0.6	2	6.6	21.5	12.6	41.5	18.6	61	24.6	80.5
0.7	2.5	6.7	22	12.7	41.5	18.7	61.5	24.7	81
0.8	2.5	6.8	22.5	12.8	42	18.8	61.5	24.8	81.5
0.9	3	6.9	22.5	12.9	42.5	18.9	62	24.9	81.5
1.0	3.5	7.0	23	13.0	42.5	19.0	62.5	25.0	82
1.1	3.5	7.1	23.5	13.1	43	19.1	62.5	25.1	82.5
1.2	4	7.2	23.5	13.2	43.5	19.2	63	25.2	82.5
1.3	4.5	7.3	24	13.3	43.5	19.3	63.5	25.3	83
1.4	4.5	7.4	24.5	13.4	44	19.4	63.5	25.4	83.5
1.5	5	7.5	24.5	13.5	44.5	19.5	64	25.5	83.5
1.6	5	7.6	25	13.6	44.5	19.6	64.5	25.6	84
1.7	5.5	7.7	25.5	13.7	45	19.7	64.5	25.7	84.5
1.8	6	7.8	25.5	13.8	45.5	19.8	65	25.8	84.5
1.9	6	7.9	26	13.9	45.5	19.9	65.5	25.9	85
2.0	6.5	8.0	26	14.0	46	20.0	65.5	26.0	85.5
2.1	7	8.1	26.5	14.1	46.5	20.1	66	26.1	85.5
2.2	7	8.2	27	14.2	46.5	20.2	66.5	26.2	86
2.3	7.5	8.3	27	14.3	47	20.3	66.5	26.3	86.5
2.4	8	8.4	27.5	14.4	47	20.4	67	26.4	86.5
2.5	8	8.5	28	14.5	47.5	20.5	67.5	26.5	87
2.6	8.5	8.6	28	14.6	48	20.6	67.5	26.6	87.5
2.7	9	8.7	28.5	14.7	48	20.7	68	26.7	87.5
2.8	9	8.8	29	14.8	48.5	20.8	68	26.8	88
2.9	9.5	8.9	29	14.9	49	20.9	68.5	26.9	88.5
3.0	10	9.0	29.5	15.0	49	21.0	69	27.0	88.5
3.1	10	9.1	30	15.1	49.5	21.1	69	27.1	89
3.2	10.5	9.2	30	15.2	50	21.2	69.5	27.2	89
3.3	11	9.3	30.5	15.3	50	21.3	70	27.3	89.5
3.4	11	9.4	31	15.4	50.5	21.4	70	27.4	90
3.5	11.5	9.5	31	15.5	51	21.5	70.5	27.5	90
3.6	12	9.6	31.5	15.6	51	21.6	71	27.6	90.5
3.7	12	9.7	32	15.7	51.5	21.7	71	27.7	91
3.8	12.5	9.8	32	15.8	52	21.8	71.5	27.8	91
3.9	13	9.9	32.5	15.9	52	21.9	72	27.9	91.5
4.0	13	10.0	33	16.0	52.5	22.0	72	28.0	92
4.1	13.5	10.1	33	16.1	53	22.1	72.5	28.1	92
4.2	14	10.2	33.5	16.2	53	22.2	73	28.2	92.5
4.3	14	10.3	34	16.3	53.5	22.3	73	28.3	93
4.4	14.5	10.4	34	16.4	54	22.4	73.5	28.4	93
4.5	15	10.5	34.5	16.5	54	22.5	74	28.5	93.5
4.6	15	10.6	35	16.6	54.5	22.6	74	28.6	94
4.7	15.5	10.7	35	16.7	55	22.7	74.5	28.7	94
4.8	15.5	10.8	35.5	16.8	55	22.8	75	28.8	94.5
4.9	16	10.9	36	16.9	55.5	22.9	75	28.9	95
5.0	16.5	11.0	36	17.0	56	23.0	75.5	29.0	95
5.1	17	11.1	36.5	17.1	56	23.1	76	29.1	95.5
5.2	17	11.2	36.5	17.2	56.5	23.2	76	29.2	96
5.3	17.5	11.3	37	17.3	57	23.3	76.5	29.3	96
5.4	17.5	11.4	37.5	17.4	57	23.4	77	29.4	96.5
5.5	18	11.5	37.5	17.5	57.5	23.5	77	29.5	97
5.6	18.5	11.6	38	17.6	57.5	23.6	77.5	29.6	97
5.7	18.5	11.7	38.5	17.7	58	23.7	78	29.7	97.5
5.8	19	11.8	38.5	17.8	58.5	23.8	78	29.8	98
5.9	19.5	11.9	39	17.9	58.5	23.9	78.5	29.9	98
6.0	19.5	12.0	39.5	18.0	59	24.0	78.5	30.0	98.5

APPENDIX H

LIST OF ACTIVE AND DISUSED WORKINGS

In late 1979 there were only two active sand and gravel quarries, both operated by Tilling Construction Services Ltd. The same company has been the operator of all the sand and gravel quarries in the district that have been worked since the Second World War.

Name and grid reference	*Approximate area (ha)*	*Remarks*
ACTIVE		
Farnham Quarry [352 596]	65*	Partly restored
Hay-a-Park Quarry [363 583]	64†	Partly restored
DISUSED		
Haughs Farm [367 563]	26	Restored
York Road, Knaresborough [363 566]	3	Restored
Low Moor Land [331 599]	4*	Unrestored
North of Crosspass House [336 600]	3*	Restored

* These figures do not include the areas of these quarries that lie outside the area of the resource sheet.

† This figure does not include workings north of Hazelheads Lane.

REFERENCES

ALLEN, V. T. 1936. Terminology of medium-grained sediments. *Rep. Natl. Res. Counc. Washington 1935–36, App. 1, Rep. Comm. Sedimentation*, 18–47.

ARCHER, A. A. 1969. Background and problems of an assessment of sand and gravel resources in the United Kingdom. *Proc. 9th Commonw. Min. Metall. Congr., 1969*, Vol. 2: Mining and petroleum geology, 495–508.

— 1970a. Standardisation of the size classification of naturally occurring particles. *Géotechnique*, Vol. 20, 103–107.

— 1970b. Making the most of metrication. *Quarry Managers' J.*, Vol. 54, No. 6, 223–227.

ATTERBERG, A. 1905. Die rationelle Klassifikation der Sande und Kiese. *Chem. Ztg.*, Vol. 29, 195–198.

BRITISH STANDARDS INSTITUTION. 1967. *Methods of testing soils for civil engineering purposes.* BS 1377: 1967. (London: British Standards Institution.)

BUREAU OF MINES AND GEOLOGICAL SURVEY. 1948. *Mineral resources of the United States.* (Washington DC: Public Affairs Press.)

BURGESS, I. C. and COOPER, A. H. 1981. The Farnham (IGS) Borehole, near Knaresborough, North Yorkshire. *Rep. Inst. Geol. Sci.*, No. 80/1, 12–17.

EDWARDS, W., MITCHELL, G. H. and WHITEHEAD, T. H. 1950. The geology of the district north and east of Leeds. *Mem. Geol. Surv. G. B.*, Sheet 70. 93 pp.

FOX-STRANGWAYS, C. 1908. The geology of the country north and east of Harrogate. *Mem. Geol. Surv. G.B.*, Sheet 62. 100 pp.

HARRIS, P. M., THURRELL, R. G., HEALING, R. A. and ARCHER, A. A. 1974. Aggregates in Britain. *Proc. R. Soc.*, Ser. A, Vol. 339, 329–353.

HARTNUP, R. 1975. Soils in North Yorkshire, II: Sheet SE 36. *Soil Surv. Rec.*, No. 30, 89 pp.

JOHNSON, P. G. 1974. Evidence for abandoned river courses in the west of the Vale of York. *Proc. Yorkshire Geol. Soc.*, Vol. 40, 223–232.

LANE, E. W. and others. 1947. Report of the sub-committee on sediment terminology. *Trans. Am. Geophys. Union*, vol. 28, 936–938.

NORTH YORKSHIRE COUNTY COUNCIL. 1978. *The Knaresborough–Staveley mineral subject-plan. Sand and gravel workings and after-use. Report of survey, issues and alternatives.* (Harrogate: North Yorkshire County Council.) 86 pp.

PETTIJOHN, F. J. 1957. *Sedimentary rocks,* 2nd edition. (London: Harper and Row.)

RAYNER, D. H. and HEMINGWAY, J. E. (Editors). 1974. The geology and mineral resources of Yorkshire. (Leeds: Yorkshire Geological Society.) 405 pp.

TERZAGHI, K. and PECK, R. B. 1967. *Soil mechanics in engineering practice,* 2nd edition. (New York: Wiley International.)

THURRELL, R. G. 1971. The assessment of mineral resources with particular reference to sand and gravel. *Quarry Managers' J.*, Vol. 55, 19–25.

TWENHOFEL, W. H. 1937. Terminology of the fine-grained mechanical sediments. *Rep. Natl. Res. Counc. Washington, 1936–37. App. 1, Rep. Comm. Sedimentation*, 81–104.

UDDEN, J. A. 1914. Mechanical composition of clastic sediments. *Bull. Geol. Soc. Am.*, Vol. 25, 655–744.

WENTWORTH, C. K. 1922. A scale of grade and class terms for clastic sediments. *J. Geol.*, Vol. 30, 377–392.

— 1935. The terminology of coarse sediments. *Bull. Natl. Res. Counc. Washington*, No. 98, 225–246.

WILLMAN, H. B. 1942. Geology and mineral resources of the Marseilles, Ottawa and Streator quadrangles. *Bull. Illinois State Geol. Surv.*, No. 66, 343–344.

The following reports of the Institute relate particularly to bulk mineral resources

Reports of the Institute of Geological Sciences

Assessment of British Sand and Gravel Resources

1 The sand and gravel resources of the country south-east of Norwich, Norfolk: Resource sheet TG 20. E. F. P. Nickless.
Report 71/20 ISBN 0 11 880216 X £1.15

2 The sand and gravel resources of the country around Witham, Essex: Resource sheet TL 81. H. J. E. Haggard.
Report 72/6 ISBN 0 11 880588 6 £1.20

3 The sand and gravel resources of the area south and west of Woodbridge, Suffolk: Resource sheet TM 24. R. Allender and S. E. Hollyer.
Report 72/9 ISBN 0 11 880596 7 £1.70

4 The sand and gravel resources of the country around Maldon, Essex: Resource sheet TL 80. J. D. Ambrose.
Report 73/1 ISBN 0 11 880600 9 £1.20.

5 The sand and gravel resources of the country around Hethersett, Norfolk: Resource sheet TG 10. E. F. P. Nickless.
Report 73/4 ISBN 0 11 880606 8 £1.60.

6 The sand and gravel resources of the country around Terling, Essex: Resource sheet TL 71. C. H. Eaton.
Report 73/5 ISBN 0 11 880608 4 £1.20

7 The sand and gravel resources of the country around Layer Breton and Tolleshunt D'Arcy, Essex: Resource sheet TL 91 and part of TL 90. J. D. Ambrose.
Report 73/8 ISBN 0 11 880614 9 £1.30

8 The sand and gravel resources of the country around Shotley and Felixstowe, Suffolk: Resource sheet TM 23. R. Allender and S. E. Hollyer.
Report 73/13 ISBN 0 11 880625 4 £1.60

9 The sand and gravel resources of the country around Attlebridge, Norfolk: Resource sheet TG 11. E. F. P. Nickless.
Report 73/15 ISBN 0 11 880658 0 £1.85

10 The sand and gravel resources of the country west of Colchester, Essex: Resource sheet TL 92. J. D. Ambrose.
Report 74/6 ISBN 0 11 880671 8 £1.45

11 The sand and gravel resources of the country around Tattingstone, Suffolk: Resource sheet TM 13. S. E. Hollyer.
Report 74/9 ISBN 0 11 880675 0 £1.95

12 The sand and gravel resources of the country around Gerrards Cross, Buckinghamshire: Resource sheet SU 99, TQ 08 and TQ 09. H. C. Squirrell.
Report 74/14 ISBN 0 11 880710 2 £2.20

Mineral Assessment Reports

13 The sand and gravel resources of the country east of Chelmsford, Essex: Resource sheet TL 70. M. R. Clarke.
ISBN 0 11 880744 7 £3.50

14 The sand and gravel resources of the country east of Colchester, Essex: Resource sheet TM 02. J. D. Ambrose.
ISBN 0 11 880745 5 £3.25

15 The sand and gravel resources of the country around Newton on Trent, Lincolnshire: Resource sheet SK 87. D. Price.
ISBN 0 11 880746 3 £3.00

16 The sand and gravel resources of the country around Braintree, Essex: Resource sheet TL 72. M. R. Clarke.
ISBN 0 11 880747 1 £3.50

17 The sand and gravel resources of the country around Besthorpe, Nottinghamshire: Resource sheet SK 86 and part of SK 76. J. R. Gozzard.
ISBN 0 11 880748 X £3.00

18 The sand and gravel resources of the Thames Valley, the country around Cricklade, Wiltshire: Resource sheet SU 09/19 and parts of SP 00/10. P. R. Robson.
ISBN 0 11 880749 8 £3.00

19 The sand and gravel resources of the country south of Gainsborough, Lincolnshire: Resource sheet SK 88 and part of SK 78. J. H. Lovell.
ISBN 0 11 880750 1 £2.50

20 The sand and gravel resources of the country east of Newark upon Trent, Nottinghamshire: Resource sheet SK 85. J. R. Gozzard
ISBN 0 11 880751 X £2.75

21 The sand and gravel resources of the Thames and Kennet Valleys, the country around Pangbourne, Berkshire: Resource sheet SU 67. H. C. Squirrell.
ISBN 0 11 880752 8 £3.25

22 The sand and gravel resources of the country north-west of Scunthorpe, Humberside: Resource sheet SE 81. J. W. C. James.
ISBN 0 11 880753 6 £3.00

23 The sand and gravel resources of the Thames Valley, the country between Lechlade and Standlake: Resource sheet SP 30 and parts of SP 20, SU 29 and SU 39. P. Robson.
ISBN 0 11 881252 1 £7.25

24 The sand and gravel resources of the country around Aldermaston, Berkshire: Resource sheet SU 56 and SU 66. H. C. Squirrell.
ISBN 0 11 881253 X £5.00

25 The celestite resources of the area north-east of Bristol: Resource sheet ST 68 and parts of ST 59, 69, 79, 58, 78, 68 and 77. E. F. P. Nickless, S. J. Booth and P. N. Mosley.
ISBN 0 11 881262 9 £5.00

26 The limestone and dolomite resources of the country around Monyash, Derbyshire: Resource sheet SK 16. F. C. Cox and D. McC. Bridge.
ISBN 0 11 881263 7 £7.00

27 The sand and gravel resources of the country west and south of Lincoln, Lincolnshire: Resource sheets SK 95, SK 96 and SK 97. I. Jackson.
ISBN 0 11 884003 7 £6.00

28 The sand and gravel resources of the country around Eynsham, Oxfordshire: Resource sheet SP 40 and part of SP 41. W. J. R. Harries.
ISBN 0 11 884012 6 £3.00

29 The sand and gravel resources of the country south-west of Scunthorpe, Humberside: Resource sheet SE 80. J. H. Lovell.
ISBN 0 11 884013 4 £3.50

30 Procedure for the assessment of limestone resources. F. C. Cox, D. McC. Bridge and J. H. Hull.
ISBN 0 11 884030 4 £1.25

31 The sand and gravel resources of the country west of Newark upon Trent, Nottinghamshire: Resource sheet SK 75. D. Price and P. J. Rogers.
ISBN 0 11 884031 2 £3.50

32 The sand and gravel resources of the country around Sonning and Henley: Resource sheet SU 77 and SU 78. H. C. Squirrell.
ISBN 0 11 884032 0 £5.25

33 The sand and gravel resources of the country north of Gainsborough: Resource sheet SK 89. J. R. Gozzard and D. Price
ISBN 0 11 884033 9 £4.50

34 The sand and gravel resources of the Dengie Peninsula, Essex: Resource sheet TL 90, etc. M. B. Simmons.
ISBN 0 11 884081 9 £5.00

35 The sand and gravel resources of the country around Darvel: Resource sheet NS 53, 63, etc. E. F. P. Nickless, A. M. Aitken and A. A. McMillan.
ISBN 0 11 884082 7 £7.00

36 The sand and gravel resources of the country around Southend-on-Sea, Essex: Resource sheets TQ 78/79 etc. S. E. Hollyer and M. B. Simmons.
ISBN 0 11 884083 5 £7.50

37 The sand and gravel resources of the country around Bawtry, South Yorkshire: Resource sheet SK 69. A. R. Clayton.
ISBN 0 11 884053 3 £5.75

38 The sand and gravel resources of the country around Abingdon, Oxfordshire: Resource sheet SU 49, 59, SP 40, 50. C. E. Corser.
ISBN 0 11 884084 5 £5.50

39 The sand and gravel resources of the Blackwater Valley (Aldershot) area: Resource sheet SU 85, 86, parts SU 84, 94, 95, 96. M. R. Clarke, A. J. Dixon and M. Kubala
ISBN 0 11 884085 1 £7.00

40 The sand and gravel resources of the country west of Darlington, County Durham: Resource sheet NZ 11, 21. A. Smtih.
ISBN 0 11 884086 X £5.00

41 The sand and gravel resources of the country around Garmouth, Grampian Region: Resource sheet NJ 36. A. M. Aitken, J. W. Merritt and A. J. Shaw.
ISBN 0 11 884090 8 £8.75

42 The sand and gravel resources of the country around Maidenhead and Marlow: Resource sheet SU 88, parts SU 87, 97, 98. P. N. Dunkley.
ISBN 0 11 884091 6 £5.00

43 The sand and gravel resources of the country around Misterton, Nottinghamshire: Resource sheet SK 79. D. Thomas and D. Price.
ISBN 0 11 884092 4 £5.25

44 The sand and gravel resources of the country around Sedgefield, Durham: Resource sheet NZ 32. M. D. A. Samuel.
ISBN 0 11 884093 2 £5.75

45 The sand and gravel resources of the country around Brampton, Cumbria: Resource sheet NY 55, part 56. I. Jackson.
ISBN 0 11 884094 0 £6.75

46 The sand and gravel resources of the country around Harlow, Essex: Resource sheet TL 41. P. M. Hopson.
ISBN 0 11 884107 6 £9.50

47 The limestone and dolomite resources of the country around Wirksworth, Derbyshire: Resource sheet SK 25, part 35. F. C. Cox and D. J. Harrison.
ISBN 0 11 884108 4 £15.00

48 The sand and gravel resources of the Loddon Valley area: Resource sheet SU 75, 76, parts 64, 65, 66 and 74. M. R. Clarke, E. J. Raynor and R. A. Sobey.
ISBN 0 11 884109 2 £8.75

49 The sand and gravel resources of the country around Lanark, Strathclyde Region: Resource sheet NS 94, part 84. J. L. Laxton and E. F. P. Nickless.
ISBN 0 11 884112 2 £11.00

50 The sand and gravel resources of the country around Fordingbridge, Hampshire: Resource sheet SU 11 and parts of SU 00, 01, 10, 20 and 21. M. Kubala.
ISBN 0 11 884111 4 £7.75

51 The sand and gravel resources of the country north of Bournemouth, Dorset: Resource sheet SU 00, 10, 20, SZ 09, 19 and 29. M. R. Clarke.
ISBN 0 11 884110 6 £9.75

52 The sand and gravel resources of the country between Hatfield Heath and Great Waltham, Essex: Resource sheet TL 51 and 61. R. J. Marks.
ISBN 0 11 884113 0 £8.00

53 The sand and gravel resources of the country around Cottenham, Cambridgeshire: Resource sheet TL 46 and 47. A. J. Dixon.
ISBN 0 11 884114 9 £9.25

54 The sand and gravel resources of the country around Huntingdon and St Ives, Cambridgeshire: Resource sheets TL 16, 17, 26, 27, 36 and 37. R. W. Gatliff.
ISBN 0 11 884115 7 £8.75

55 The sand and gravel resources of the country around Ipswich, Suffolk: Resource sheet TM 14. R. Allender and S. E. Hollyer.
ISBN 0 11 884116 5 £10.00

56 Procedure for the assessment of the conglomerate resources of the Sherwood Sandstone Group. D. P. Piper and P. J. Rogers.
ISBN 0 11 884143 2 £1.25

57 The conglomerate resources of the Sherwood Sandstone Group of the country around Cheadle, Staffordshire: Resource sheet SK 04. P. J. Rogers, D. P. Piper and T. J. Charsley.
ISBN 0 11 884144 0 *not yet priced*

58 The sand and gravel resources of the country west of Peterhead, Grampian Region: Resource sheet NK 04 and parts of NJ 94 and 95, NK 05, 14 and 15. A. A. McMillan and A. M. Aitken.
ISBN 0 11 884145 9 *not yet priced*

59 The sand and gravel resources of the country around Newbury, Berkshire: Resource sheets SU 46 and 57, parts of SU 36, 37 and 47. J. R. Gozzard.
ISBN 0 11 884146 7 *not yet priced*

60 The sand and gravel resources of the country south-west of Peterborough, in Cambridgeshire and east Northamptonshire: Resource sheets TL 09 and 19, and SP 98 and TL 08. A. M. Harrisson.
ISBN 0 11 884147 5 £15.50

61 The sand and gravel resources of the country north of Wrexham, Clwyd: Resource sheet SJ 35 and part of SJ 25. P. N. Dunkley.
ISBN 0 11 884148 3 *not yet priced*

62 The sand and gravel resources of the country around Dolphinton, Strathclyde Region, and West Linton, Borders Region: Resource sheets NT 04 and 14, and parts of NT 05 and 15. A. A. McMillan, J. L. Laxton and A. J. Shaw.
ISBN 0 11 884149 1 £8.00

63 The sand and gravel resources of the country of the valley of the Douglas Water, Strathclyde: Resource sheet NS 83 and parts of NS 82, 92 and 93. A. J. Shaw and E. F. P. Nickless
ISBN 0 11 884150 5 £11.50

64 The sand and gravel resources of the country between Wallingford and Goring, Oxfordshire: Resource sheet SU 68 and part SU 58. C. E. Corser.
ISBN 0 11 884151 3 *not yet priced*

65 The sand and gravel resources of the country around Hexham, Northumberland: Resource sheet NY 86 and 96. J. H. Lovell
ISBN 0 11 884152 1 £7.50

66 The sand and gravel resources of the country west of Chelmsford, Essex: Resource sheet TL 60. P. M. Hopson.
ISBN 0 11 884153 X £8.50

67 The sand and gravel resources of the country around Hatfield and Cheshunt, Hertfordshire: Resource sheet TL 20 and 30, and parts of TQ 29 and 39. J. R. Gozzard.
ISBN 0 11 884167 X £10.00

68 The sand and gravel resources of the country north-east of Halstead, Essex: Resource sheet TL 83. R. J. Marks and J. W. Merritt.
ISBN 0 11 884168 8 £13.25

69 The sand and gravel resources of the country around Welwyn Garden City, Hertfordshire: Resource sheet TL 11 and 21. J. R. Gozzard.
ISBN 0 11 884169 6 £10.50

70 The sand and gravel resources of the country east of Harrogate, North Yorkshire: Resource sheet SE 35. D. L. Dundas.
ISBN 0 11 884170 X £15.50

Reports of the Institute of Geological Sciences

Other Reports

69/9 Sand and gravel resources of the inner Moray Firth. A. L. Harrison and J. D. Peacock.
ISBN 0 11 880106 6 35p

70/4 Sands and gravels of the southern counties of Scotland. G. A. Goodlet.
ISBN 0 11 880105 8 90p

72/8 The use and resources of moulding sand in Northern Ireland. R. A. Old.
ISBN 0 11 881594 0 30p

73/9 The superficial deposits of the Firth of Clyde and its sea lochs. C. E. Deegan, R. Kirby, I. Rae and R. Floyd.
ISBN 0 11 880617 3 95p

77/1 Sources of aggregate in Northern Ireland (2nd edition). I. B. Cameron.
ISBN 0 11 881279 3 70p

77/2 Sand and gravel resources of the Grampian Region. J. D. Peacock and others.
ISBN 0 11 881282 3 80p

77/5 Sand and gravel resources of the Fife Region. M. A. E. Browne.
ISBN 0 11 884004 5 60p

77/6 Sand and gravel resources of the Tayside Region. I. B. Paterson.
ISBN 0 11 884008 8 £1.40

77/8 Sand and gravel resources of the Strathclyde Region. I. B. Cameron and others.
ISBN 0 11 884028 2 £2.50

77/9 Sand and gravel resources of the Central Region, Scotland. M. A. E. Browne.
ISBN 0 11 884016 9 £1.35

77/19 Sand and gravel resources of the Borders Region, Scotland. A. D. McAdam.
ISBN 0 11 884025 8 £1.00

77/22 Sand and gravel resources of the Dumfries and Galloway Region of Scotland. I. B. Cameron.
ISBN 0 11 884021 5 £1.20

78/1 Sand and gravels of the Lothian Region of Scotland. A. D. McAdam.
ISBN 0 11 884042 8 £1.00

78/8 Sand and gravel resources of the Highland Region. W. Mykura, D. L. Ross and F. May.
ISBN 0 11 884050 9 £3.00

Dd 696521 K8

Typeset for the Institute of Geological Sciences
by John Wright and Sons (Printing) Ltd, Bristol

Printed in England for Her Majesty's Stationery Office
by Commercial Colour Press, London E7